5 STEPS TO A 5™

AP Physics B

2014 EDITION

5 STEPS TO A 5™

AP Physics B

2014 EDITION

Greg Jacobs • Joshua Schulman

New York Chicago San Francisco Athens London Madrid
Mexico City Milan New Delhi Singapore Sydney Toronto

1 2 3 4 5 6 7 8 9 10 11 12 13 14 15 16 17 18 19 20 21 22 23 24   DOH/DOH   1 0 9 8 7 6 5 4 3

ISBN 978-0-07-180299-4
MHID   0-07-180299-1

e-ISBN 978-0-07-180300-7
e-MHID   0-07-180300-9
ISSN   1949-1646

McGraw-Hill Education, the McGraw-Hill Education logo, *5 Steps to a 5*, and related trade dress are trademarks or registered trademarks of McGraw-Hill Education and/or its affiliates in the United States and other countries and may not be used without written permission. All other trademarks are the property of their respective owners. McGraw-Hill Education is not associated with any product or vendor mentioned in this book.

AP, Advanced Placement Program, and *College Board* are registered trademarks of the College Entrance Examination Board, which was not involved in the production of, and does not endorse, this product.

The series editor was Grace Freedson, and the project editor was Del Franz.
Series design by Jane Tenenbaum.

McGraw-Hill Education products are available at special quantity discounts to use as premiums and sales promotions or for use in corporate training programs. To contact a representative, please visit the Contact Us pages at www.mhprofessional.com.

This book is printed on acid-free paper.

CONTENTS

STEP 4   **Review the Knowledge You Need to Score High**

STEP 5

Build Your Test-Taking Confidence

Appendices

ACKNOWLEDGMENTS

JS: I would never have written this book without the inspiration from my AP Physics C classmates—Justin Kreindel, Zack Menegakis, Adam Ohren, Jason Sheikh, and Joe Thistle—fellow physics enthusiasts who are also some of my closest friends. They were the ones who convinced me that learning physics could be a process filled not just with problem sets and formulas, but also with light-hearted antics and belly-aching laughter. If this book is fun to read, it is because of their influence.

We also extend our thanks to Grace Freedson, who was the driving force behind this book's publication, and especially to Don Reis, who was not only a superb editor but also an unwavering source of support.

GJ: Josh has already mentioned Don, Grace, and our 1999–2000 Physics C gang. I appreciate them, too. I also appreciate Ruth Mills's awesome work on the second edition and Bev Weiler and Clara Wente's careful and thorough editing of the questions and example problems for the 2010–2011 edition.

Thank-you to Chat Hull and Jessica Broaddus, veterans of my 2002 Physics B class, who provided the idea for two free-response questions.

My 2004 classes at Woodberry Forest School were extremely helpful in the development of this book. It was they who served as guinea pigs, reading chapters for clarity and correctness, making suggestions, and finding mistakes. They are Andrew Burns, Jordan Crittenden, David Fulton, Henry Holderness, Michael Ledwith, Johnny Phillips, Rob Sellers, and Chris Straka from Physics C; Wes Abcouwer, Wyatt Bone, Matt Brown, David Goodson, Bret Holbrook, Mike Johnson, Rich Lane, Jake Miller, Jake Reeder, Charles Shackelford, Beau Thomas, David Badham, Marks Brewbaker, Charlton deSaussure, Palmer Heenan, Wilson Kieffer, Brian McCormick, Eli Montague, Christian Rizzuti, Pierre Rodriguez, and Frazier Stowers from Physics B; and Andy Juc, Jamie Taliaferro, Nathan Toms, Matt Laughridge, Jamie Gardiner, Graham Gardiner, Robbie Battle, William Crosscup, Jonas Park, Billy Butler, Bryan May, Fletcher Fortune, and Stuart Coleman from the general physics class. Although Josh and I bear responsibility for all errors in the text, the folks mentioned above deserve credit for minimizing our mistakes.

The idea for the 4-minute drill came originally from Keen Johnson Babbage, my seventh-grade social studies teacher. I've borrowed the idea from him for 16 years of teaching AP. Thank you!

The faculty and administration at Woodberry, in particular Jim Reid, the science department chairman, deserve mention. They have been so supportive of me professionally.

Additional thanks go to members of my 2009 AP physics classes who helped edit the practice tests: Min SuKim, Cannon Allen, Collins MacDonald, Luke Garrison, Chris Cirenza, and Landon Biggs.

Most important, I'd like to thank Shari and Milo Cebu for putting up with me during all of my writing projects.

ABOUT THE AUTHORS

GREG JACOBS teaches both the B and C levels of AP physics at Woodberry Forest School, the nation's premier boarding school for boys. He is a reader and consultant for the College Board—this means he grades the AP Physics exams, and he runs professional development seminars for other AP teachers. Greg is president of the USAYPT, a non-profit organization promoting physics research at the high school level. Greg was recently honoured as an AP Teacher of the Year by the Siemens Foundation. Outside the classroom, Greg is head debate coach for Woodberry Forest, and he umpires high school baseball. He and faculty member Peter Cashwell broadcast Woodberry varsity baseball and football over the Internet. Greg writes a physics teaching blog available at **www.jacobsphysics.blogspot.com**.

JOSH SCHULMAN earned 5s on both the AP Physics B and AP Physics C exams in high school and was chosen as a semifinalist in the selection of the U.S. Physics Team. Josh graduated from Princeton University in 2004 with a Bachelor's degree in chemistry. He graduated from Harvard Medical School in 2009.

INTRODUCTION: THE FIVE-STEP PROGRAM

Welcome!

We know that preparing for the Advanced Placement (AP) Physics B exam can seem like a daunting task. There's a lot of material to learn, and some of it can be rather challenging. But we also know that preparing for the AP exam is much easier—and much more enjoyable—if you do it with a friendly guide. So let us introduce ourselves: our names are Josh and Greg, and we'll be your friendly guides for this journey.

Why This Book?

To understand what makes this book unique, you should first know a little bit about who we are. Josh took both the AP Physics B and AP Physics C exams during high school. In fact, Josh started writing this book soon after graduating from high school, when the experience of studying for these exams was fresh in his memory. Greg has taught AP Physics B and C for more than 15 years, helping more than 70% of his students garner 5s on the exam. He is also an AP Physics table leader—which means he sets the rubrics for the AP exams and supervises their scoring.

In other words, we offer a different perspective from that of most other review books. Between the two of us, we know firsthand what it's like to prepare for the AP exam, and we know firsthand what it takes to get a good score on the AP exam.

We also know, from our own experiences and from talking with countless other students and teachers, what you *don't* need in a review book. You don't need to be overwhelmed with unimportant, technical details; you don't need to read confusing explanations of arcane topics; you don't need to be bored with a dull text.

Instead, what we think you do need—and what this book provides—are the following:

- A text that's written in clear, simple language.
- A thorough review of every topic you need to know for the AP exam.
- Lots of problem-solving tips and strategies.
- An introduction to the student-tested Five-Step Program to Mastering the AP Physics B Exam.

Organization of the Book: The Five-Step Program

You will be taking a lengthy, comprehensive exam this May. You want to be well prepared enough that the exam takes on the feel of a command performance, not a trial by fire. Following the Five-Step program is the best way to structure your preparation.

Step 1: Set Up Your Study Program

Physics does not lend itself well to cramming. Success on the AP exam is invariably the result of diligent practice over the course of months, not the result of an all-nighter on the eve of exam day. Step 1 gives you the background and structure you need before you even start exam preparation.

Step 2: Determine Your Test Readiness

We've included a diagnostic test, of course, broken down by topic. But more important to your preparation are the *fundamentals quizzes* in Chapter 4. These quizzes, a unique feature of the *5 Steps to a 5* program, are different from test-style problems.

A problem on the AP exam usually requires considerable problem solving or critical thinking skills. Rare is the AP question that asks about straightforward facts that you can memorize—you'll get maybe two of those on the entire 70-question multiple-choice test. Rather than asking you to spit out facts, the AP exam asks you to use the facts you know to reason deeply about a physical situation. But if you don't know the fundamental facts, you certainly won't be able to reason deeply about anything!

Thus, a good place to start your test preparation is by quizzing yourself. Find out what fundamental facts you know, and which you need to know. The *5 Steps* fundamentals quizzes will diagnose your areas of strength and weakness. Once you can answer every question on a fundamentals quiz quickly and accurately, you are ready for deeper questions that will challenge you on the AP exam.

Step 3: Develop Strategies for Success

Yes, yes, we know you've been listening to general test-taking advice for most of your life. Yet, we have *physics-specific* advice for you. An AP physics test requires a dramatically different approach than does a state standards test or an SAT.

We start you with the secret weapon in attacking an AP test: memorizing equations. We explain *why* you should memorize, then we suggest some ways to make the learning process smoother. Next, we move on to discuss the major types of questions you'll see on the AP exam, and how to approach each with confidence.

Finally, we present you with drills on some of the most common physics situations tested on the AP exams. These exercises will allow you to conquer any fear or uncertainty you may have about your skills.

Step 4: Review the Knowledge You Need to Score High

This is a comprehensive review of all the topics on the AP exam. Now, you've probably been in an AP Physics class all year; you've likely read[1] your textbook. Our review is meant to be just that—*review*, in a readable format, and focused exclusively on the AP exam.

These review chapters are appropriate both for quick skimming, to remind yourself of salient points, and for in-depth study, working through each practice problem. We do not go into nearly as much detail as a standard textbook; but the advantage of our lack of detail is that we can focus only on those issues germane to the AP Physics B exam.

Step 5: Build Your Test-Taking Confidence

Here is your full-length practice test. Unlike other practice tests you may take, this one comes with thorough explanations. One of the most important elements in learning physics

---

[1] Or at least tried to read.

is making, and then learning from, mistakes. We don't just tell you what you got wrong; we explain why your answer is wrong, and how to do the problem correctly. It's okay to make a mistake here, because if you do, you won't make that same mistake again on that Monday in mid-May.

The Graphics Used in this Book

To emphasize particular skills and strategies, we use several icons throughout this book. An icon in the margin will alert you that you should pay particular attention to the accompanying text. We use these three icons:

1. This icon points out a very important concept or fact that you should not pass over.

2. This icon calls your attention to a problem-solving strategy that you may want to try.

3. This icon indicates a tip that you might find useful.

Boldfaced words indicate terms that are included in the glossary at the end of the book. Boldface is also used to indicate the answer to a sample problem discussed in the test.

5 STEPS TO A 5™

AP Physics B

2014 EDITION

Set Up Your Study Program

CHAPTER 1

How to Approach Your AP Physics Course

IN THIS CHAPTER

Summary: Recognize the difference between truly understanding physics and just doing well in your physics class.

Key Ideas

✪ Focus on increasing your knowledge of physics, not on pleasing your teacher.

✪ Don't spend more than 10 minutes at one time on a problem without getting anywhere—come back to it later if you don't get it.

✪ Form a study group; your classmates can help you learn physics better.

✪ If you don't understand something, ask your teacher for help.

✪ Don't cram; although you can memorize equations, the skills you need to solve physics problems can't be learned overnight.

Before we even dive into the nitty-gritty of the AP Physics B exam, it's important for you to know that the AP exam is an *authentic* physics test. What this means is that it's not possible to "game" this test—in order to do well, *you must know your physics*. Therefore, the purpose of this book is twofold:

(1) to teach you the ways in which the AP exam tests your physics knowledge, and
(2) to give you a review of the physics topics that will be tested—and to give you some hints on how to approach these topics.

Everyone who takes the AP exam has just completed an AP Physics course. **Recognize that your physics course is the place to start your exam preparation!** Whether or not you are satisfied with the quality of your course or your teacher, the best way to start preparing for the exam is by doing careful, attentive work in class all year long.

Okay, for many readers, we're preaching to the choir. You don't want to hear about your physics class; you want the specifics about the AP exam. If that's the case, go ahead and turn to Chapter 2, and get started on your exam-specific preparation. But we think that you can get even more out of your physics class than you think you can. Read these pieces of time-tested advice, follow them, and we promise you'll feel more comfortable about your class *and* about the AP exam.

Ignore Your Grade

This must be the most ridiculous statement you've ever read. But it may also be the most important of these suggestions. Never ask yourself or your teacher "Can I have more points on this assignment?" or "Is this going to be on the test?" You'll worry so much about giving the teacher merely what she or he wants that you won't learn physics in the way that's best for you. Whether your score is perfect or near zero, ask, "Did I really understand all aspects of these problems?"

Remember, the AP exam tests your physics knowledge. If you understand physics thoroughly, you will have no trouble at all on the AP test. But, while you may be able to argue yourself a better grade in your physics *class*, even if your comprehension is poor, the AP readers are not so easily moved.

If you take this advice—if you really, truly ignore your grade and focus on physics— your grade will come out in the wash. You'll find that you got a very good grade after all, because you understood the subject so well. But you *won't care*, because you're not worried about your grade!

Don't Bang Your Head Against a Brick Wall

Our meaning here is figurative, although there are literal benefits also. Never spend more than 10 minutes or so staring at a problem without getting somewhere. If you honestly have no idea what to do at some stage of a problem, STOP. Put the problem away. Physics has a way of becoming clearer after you take a break.

On the same note, if you're stuck on some algebra, don't spend forever trying to find what you know is a trivial mistake, say a missing negative sign or some such thing. Put the problem away, come back in an hour, and start from scratch. This will save you time in the long run.

And finally, if you've put forth a real effort, you've come back to the problem many times and you still can't get it: relax. Ask the teacher for the solution, and allow yourself to be enlightened. You will not get a perfect score on every problem. But you don't care about your grade, remember?

Work with Other People

When you put a difficult problem aside for a while, it always helps to discuss the problem with others. Form study groups. Have a buddy in class with whom you are consistently comparing solutions.

Although you may be able to do all your work in every other class without help, we have never met a student who is capable of solving every physics problem on his or her own.

It is not shameful to ask for help. Nor is it dishonest to seek assistance—as long as you're not copying, or allowing a friend to carry you through the course. Group study is permitted and encouraged in virtually every physics class around the globe.

Ask Questions When Appropriate

We know your physics teacher may seem mean or unapproachable, but in reality, physics teachers do want to help you understand their subject. If you don't understand something, don't be afraid to ask. Chances are that the rest of the class has the same question. If your question is too basic or requires too much class time to answer, the teacher will tell you so.

Sometimes the teacher will not answer you directly, but will give you a hint, something to think about so that you might guide yourself to your own answer. Don't interpret this as a refusal to answer your question. You must learn to think for yourself, and your teacher is helping you develop the analytical skills you need for success in physics.

Keep an Even Temper

A football team should not give up because they allow an early field goal. Similarly, you should not get upset at poor performance on a test or problem set. No one expects you to be perfect. Learn from your mistakes, and move on—it's too long a school year to let a single physics assignment affect your emotional state.

On the same note, however, a football team should not celebrate victory because it scores a first-quarter touchdown. You might have done well on this test, but there's the rest of a nine-month course to go. Congratulate yourself, then concentrate on the next assignment.

Don't Cram

Yes, we know that you got an "A" on your history final because, after you slept through class all semester, you studied for 15 straight hours the day before the test and learned everything. And, yes, we know you are willing to do the same thing this year for physics. We warn you, both from our and from others' experience: *it won't work*. Physics is not about memorization and regurgitation. Sure, there are some equations you need to memorize. But problem-solving skills cannot be learned overnight.

Furthermore, physics is cumulative. The topics you discuss in December rely on the principles you learned in September. If you don't understand basic vector analysis and free-body diagrams, how can you understand the relationship between an electric field (which is a vector quantity) and an electric force, or the multitude of other vector quantities that you will eventually study?

So, the answer is to keep up with the course. Spend some time on physics every night, even if that time is only a couple of minutes, even if you have no assignment due the next day. Spread your "cram time" over the entire semester.

Never Forget, Physics Is "Phun"

The purpose of all these problems, these equations, and these exams is to gain knowledge about physics—a deeper understanding of how the natural world works. Don't be so caught up in the grind of your coursework that you fail to say "Wow!" occasionally. Some of the things you're learning are truly amazing. Physics gives insight into some of humankind's most critical discoveries, our most powerful inventions, and our most fundamental technologies. Enjoy yourself. You have an opportunity to emerge from your physics course with wonderful and useful knowledge, and unparalleled intellectual insight. Do it.

CHAPTER 2

What You Need to Know About the AP Physics B Exam

IN THIS CHAPTER

Summary: Learn what topics are tested, how the test is scored, and basic test-taking information.

Key Ideas

✪ Most colleges will award credit for a score of 4 or 5, some for a 3.

✪ Multiple-choice questions account for half of your final score.

✪ There is no penalty for guessing on the multiple-choice questions. You should answer every question.

✪ Free-response questions account for half of your final score.

✪ Your composite score on the two test sections is converted to a score on the 1-to-5 scale.

Background Information

The AP Physics B exam that you will take was first offered by the College Board in 1954. Since then, the number of students taking the test has grown rapidly. In 2009, more than 50,000 students took the AP Physics B exam. By 2011, the number of exam takers had risen to more than 60,000. The number of students taking the AP Physics B exam is expected to continue to increase.

Some Frequently Asked Questions About the AP Physics B Exam

Why Should I Take the AP Physics Exam?

Many of you take the AP Physics exam because you are seeking college credit. The majority of colleges and universities will award you some sort of credit for scoring a 4 or a 5. A smaller number of schools will even accept a 3 on the exam. This means you are one or two courses closer to graduation before you even start college!

Therefore, one compelling reason to take the AP exam is economic. How much does a college course cost, even at a relatively inexpensive school? You're talking several thousand dollars. If you can save those thousands of dollars by paying less than a hundred dollars now, why not do so?

Even if you do not score high enough to earn college credit, the fact that you elected to enroll in AP courses tells admission committees that you are a high achiever and serious about your education. In recent years, about 60% of students have scored a 3 or higher on their AP Physics B or C exam.

You'll hear a whole lot of misinformation about AP credit policies. Don't believe anything a friend (or even an adult) tells you; instead, find out for yourself. A good way to learn about the AP credit policy of the school you're interested in is to look it up on the College Board's official Web site, at http://collegesearch.collegeboard.com/apcreditpolicy/index.jsp. Even better, contact the registrar's office or the physics department chairman at the college directly.

Isn't the AP Physics B Exam Changing Soon?

Yes. While the 2014 administration of the Physics B exam will be exactly the same as it's been for years, in May 2015, Physics B will no longer exist. In its place, the College Board has created two new exams called "AP Physics 1" and "AP Physics 2." These exams will still test introductory algebra-based physics, but the emphasis of the tests will shift ever farther away from calculation and toward verbal explanation of the relevant physics principles.

But, *so what?* If you've bought this book, presumably you are intending to take the AP Physics exam in May 2014. This book prepares you for the style of exam that has been administered for the last couple of decades. Your teacher may be anxious about how the AP exam changes will affect your school's physics classes, but there's no need for you to care about the new exams, any more than you worry about the fact that the sun only has about 5 billion years' worth of fuel left: yes, it's important, but it won't affect YOU in any way.

What's the Difference Between Physics B and Physics C?

There are two AP Physics courses that you can take—Physics B and Physics C—and they differ in both the range of topics covered and the level at which those topics are tested. Here's the rundown. This book is specifically designed to prepare you for AP Physics B. If you decide to attempt the Physics C exam, try *5 Steps to a 5: AP Physics C*.

Physics B

As a survey course, Physics B covers a broad range of topics. This book's table of contents lists them all. This course is algebra-based—no calculus is necessary. In fact, the most difficult math you will encounter is solving two simultaneous algebraic equations, which is probably something you did in ninth grade.

The B course is ideal for ALL college-bound high school students. For those who intend to major in math or the heavy duty sciences, Physics B serves as a perfect introduction to college-level work. For those who want nothing to do with physics after high school, Physics B is a terrific terminal course—you get exposure to many facets of physics at a rigorous yet understandable level.

Most importantly, for those who aren't sure in which direction their college career may head, the B course can help you decide: "Do I like this stuff enough to keep studying it, or not?"

Although it is intended to be a second-year course, Physics B is often successfully taught as an intensive first-time introduction to physics.

Physics C

These courses are ONLY for those who have already taken a solid introductory physics course and are considering a career in math or science. Some schools teach Physics C as a follow-up to Physics B, but as long as you've had a rigorous introduction to the subject, that introduction does not have to be at the AP level.

Physics C is two separate courses: (1) Newtonian Mechanics, and (2) Electricity and Magnetism. Of course, the B course covers these topics as well. However, the C courses go into greater depth and detail. The problems are more involved, and they demand a higher level of conceptual understanding. You can take either or both 90-minute Physics C exams.

The C courses require some calculus. Although much of the material can be handled without it, you should be taking a good calculus course concurrently.

Is One Exam Better than the Other? Should I Take Them Both?

We strongly recommend taking only one exam—the one your high school AP course prepared you for. Physics C is not considered "better" than Physics B in the eyes of colleges and scholarship committees. They are different courses with different intended audiences. It is far better to do well on the one exam you prepared for than to do poorly on both exams.

What Is the Format of the Exam?

Table 1.1 summarizes the format of the AP Physics B exam.

Table 1.1   AP Physics B exam

AP Physics B

SECTION	NUMBER OF QUESTIONS	TIME LIMIT
I. Multiple-Choice Questions	70	1 hour and 30 minutes
II. Free-Response Questions	6–8, depending on the year	1 hour and 30 minutes

Who Writes the AP Physics Exam?

Development of each AP exam is a multiyear effort that involves many education and testing professionals and students. At the heart of the effort is the AP Physics Development Committee, a group of college and high-school physics teachers who are typically asked to serve for three years. The committee and other physics teachers create a large pool of multiple-choice questions. With the help of the testing experts at Educational Testing Service (ETS), these questions are then pre-tested with college students for accuracy, appropriateness, clarity, and assurance that there is only one possible answer. The results of this pre-testing allow each question to be categorized by degree of difficulty. After several more months of development and refinement, Section I of the exam is ready to be administered.

The free-response questions that make up Section II go through a similar process of creation, modification, pre-testing, and final refinement so that the questions cover the necessary areas of material and are at an appropriate level of difficulty and clarity. The committee also makes a great effort to construct a free-response exam that will allow for clear and equitable grading by the AP readers.

At the conclusion of each AP reading and scoring of exams, the exam itself and the results are thoroughly evaluated by the committee and by ETS. In this way, the College Board can use the results to make suggestions for course development in high schools and to plan future exams.

What Topics Appear on the Exam?

The College Board, after consulting with physics teachers at all levels, develops a curriculum that covers material that college professors expect to cover in their first-year classes. Based on this outline of topics, the multiple-choice exams are written such that those topics are covered in proportion to their importance to the expected understanding of the student.

Confused? Suppose that faculty consultants agree that, say, atomic and nuclear physics is important to the physics curriculum, maybe to the tune of 10%. If 10% of the curriculum is devoted to atomic and nuclear physics, then you can expect roughly 10% of the exam will address atomic and nuclear physics. This includes both the multiple-choice and the free-response sections—so a topic that is not tested in the free-response section will have *extra* multiple-choice questions to make up the difference.

The following is the general outline for the AP Physics B curriculum and exam. Remember this is just a guide, and each year the exam differs slightly in the percentages.

AP PHYSICS B

I.	Newtonian Mechanics	35%
II.	Fluid Mechanics and Thermal Physics	15%
III.	Electricity and Magnetism	25%
IV.	Waves and Optics	15%
V.	Atomic and Nuclear Physics	10%

What Types of Questions Are Asked on the Exam?

The multiple-choice questions tend to focus either on your understanding of concepts or on your mastery of equations and their meaning. Here's an example of a "concept" multiple-choice question.

Which of the following is an expression of conservation of charge?

(A) Kirchoff's loop rule
(B) Kirchoff's junction rule
(C) Ohm's law
(D) Snell's law
(E) Kinetic theory of gases

The answer is **B**. Kirchoff's junction rule states that whatever charge comes in must come out. If you don't remember Kirchoff's junction rule, turn to Chapter 20, Circuits.

And here's an example of an "equation" multiple-choice question.

If the separation between plates in a parallel-plate capacitor is tripled, what happens to the capacitance?

(A) It is reduced by a factor of 9.
(B) It is reduced by a factor of 3.
(C) It remains the same.
(D) It increases by a factor of 3.
(E) It increases by a factor of 9.

The answer is **B**. For this kind of question, you either remember the equation for the capacitance of a parallel plate capacitor,

$$C = \frac{\varepsilon_0 A}{d}$$

or you don't. For help, turn to Chapter 6, "Memorizing Equations in the Shower."

For the multiple-choice part of the exam, you are given a sheet that contains a bunch of physical constants (like the mass of a proton), SI units, and trigonometric values (like "tan 45° = 1"). All in all, this sheet is pretty useless—you'll probably only refer to it during the course of the test if you need to look up an obscure constant. That doesn't happen as often as you might think.

The free-response questions take 10–15 minutes apiece to answer, and they test both your understanding of concepts and your mastery of equations. Some of the free-response questions ask you to design or interpret the results of an experimental setup; others are more theoretical. Luckily, for this portion of the exam, in addition to the constant sheet you get with the multiple-choice questions, you will also get a sheet that contains every equation you will ever need.

We talk in much more detail about both the multiple-choice and the free-response sections of the test later, in Step 5, so don't worry if this is all a bit overwhelming now.

Who Grades My AP Physics Exam?

Every June, a group of physics teachers gathers for a week to assign grades to your hard work. Each of these "readers" spends a day or so getting trained on one question—and one question only. Because each reader becomes an expert on that question, and because each exam book is anonymous, this process provides a very consistent and unbiased scoring of that question.

During a typical day of grading, a random sample of each reader's scores is selected and cross-checked by other experienced "Table Leaders" to ensure that the consistency is maintained throughout the day and the week. Each reader's scores on a given question are also statistically analyzed, to make sure they are not giving scores that are significantly higher or lower than the mean scores given by other readers of that question. All measures are taken to maintain consistency and fairness for your benefit.

Will My Exam Remain Anonymous?

Absolutely. Even if your high-school teacher happens to randomly read your booklet, there is virtually no way he or she will know it is you. To the reader, each student is a number, and to the computer, each student is a bar code.

What About That Permission Box on the Back?

The College Board uses some exams to help train high-school teachers so that they can help the next generation of physics students to avoid common mistakes. If you check this box, you simply give permission to use your exam in this way. Even if you give permission, your anonymity is still maintained.

How Is My Multiple-Choice Section Scored?

The multiple-choice section of each physics exam is worth half of your final score. Your answer sheet is run through the computer, which adds up your correct responses. The number of correct responses is your raw score on the multiple-choice section.

If I Don't Know the Answer, Should I Guess?

Yes. There is no penalty for guessing.

How Is My Free-Response Section Scored?

Your performance on the free-response section is also worth half of your final score. The Physics B free-response section will consist of longer questions, worth 15 points, and slightly shorter questions, worth 10 points. Your score on the free-response section is simply the sum of your scores on each problem.

How Is My Final Grade Determined and What Does It Mean?

Each section counts for 50% of the exam. The total composite score is thus a weighted sum of the multiple-choice and the free-response sections. In the end, when all of the numbers have been crunched, the Chief Faculty Consultant converts the range of composite scores to the 5-point scale of the AP grades. This conversion is not a true curve—it's not that there's some target percentage of 5s to give out. This means you're not competing against other test takers. Rather, the 5-point scale is adjusted each year to reflect the same standards as in previous years. The goal is that students who earn 5s this year are just as strong as those who earned 5s in 2000 or 2005.

The tables at the end of the practice exams in this book give you a rough example of a conversion, and as you complete the practice exams, you should use this to give yourself a hypothetical grade. Keep in mind that the conversion changes slightly every year to adjust for the difficulty of the questions—but, generally, it takes only about 65% of the available points to earn a 5.

You should receive your grade in early July.

How Do I Register and How Much Does It Cost?

If you are enrolled in AP Physics in your high school, your teacher will provide all of these details, but a quick summary here can't hurt. After all, you do not have to enroll in the AP course to register for and complete the AP exam. When in doubt, the best source of information is the College Board's Web site: www.collegeboard.com.

In 2012, the fee for taking the exam was $87. Students who demonstrate financial need may receive a refund to offset the cost of testing. The fee and the refund usually change a little from year to year. You can find out more about the exam fee and fee reductions and subsidies from the coordinator of your AP program or by checking information on the official Web site: www.collegeboard.com.

I know that seems like a lot of money just for a test. But, you should think of this $87 as the biggest bargain you'll ever find. Why? Most colleges will give you a few credit hours for a good score. Do you think you can find a college that offers those credit hours for less than $87? Usually you're talking hundreds of dollars per credit hour! You're probably saving thousands of dollars by earning credits via AP.

There are also several optional fees that must be paid if you want your scores rushed to you or if you wish to receive multiple-grade reports. Don't worry about doing that unless your college demands it. (What, you think your scores are going to change if you don't find them out right away?)

The coordinator of the AP program at your school will inform you where and when you will take the exam. If you live in a small community, your exam may not be administered at your school, so be sure to get this information.

What If My School Only Offers AP Physics B and Not AP Physics C, or Vice Versa? Or, What If My School Doesn't Offer AP Physics at All?

Ideally, you should enroll in the AP class for the exam you wish to take. But, not every school offers exactly what you want to take.

If your school offers one exam or the other, you are much better off taking the exam for which your teacher prepared you. Sure, if you are an absolute top Physics B student, you can probably pass the Physics C exam with some extra preparation; but if you're a top Physics B student, why not just earn your 5 on the B exam rather than take a chance at merely passing the C exam? Or, if you've been preparing for Physics C, you might think you have a better chance for success on the "easier" B exam. But, the B exam tests all sorts of topics that aren't covered in Physics C, so you're still most likely better off on the exam your class taught toward.

If your school doesn't offer either AP course, then you should look at the content outline and talk to your teacher. Chances are, you will want to take the B exam, and chances are you will have to do a good bit of independent work to learn the topics that your class didn't discuss. But, if you are a diligent student in a rigorous course, you will probably be able to do fine.

What Should I Bring to the Exam?

On exam day, I suggest bringing the following items:

- Several pencils and an eraser that doesn't leave smudges.
- Black or blue colored pens for the free-response section.[1]
- A ruler or straightedge.
- A scientific calculator with fresh batteries. (A graphing calculator is not necessary.)
- A watch so that you can monitor your time. You never know if the exam room will have a clock on the wall. Make sure you turn off the beep that goes off on the hour.
- Your school code.
- Your photo identification and social security number.
- Tissues.
- Your quiet confidence that you are prepared.

---

[1]You may use a pencil, but there's no need . . . you should not erase incorrect work, you should cross it out. Not only does crossing out take less time than erasing, but if you erase by mistake, you lose all your work. But, if you change your mind about crossing something out, just circle your work and write the reader a note: "Grade this!"

What Should I NOT Bring to the Exam?

Leave the following at home:

- A cell phone, PDA, or walkie-talkie.
- Books, a dictionary, study notes, flash cards, highlighting pens, correction fluid, etc., *including this book.* Study aids won't help you the morning of the exam . . . end your studying in the very early evening the night before.
- Portable music of any kind. No iPods, MP3 players, or CD players.
- Clothing with any physics terminology or equations on it.
- Panic or fear. It's natural to be nervous, but you can comfort yourself that you have used this book well and that there is no room for fear on your exam.

CHAPTER 3

How to Plan Your Time

IN THIS CHAPTER

Summary: What to study for the Physics B exam, plus three schedules to help you plan.

Key Ideas
✪ Focus your attention and study time on those topics that are most likely to increase your score.
✪ Study the topics that you're *afraid* will appear, and relax about those that you're best at.
✪ Don't study so widely that you don't get good at some specific type of problem.

The AP Physics exam is held on a Monday afternoon in mid-May. You may think that you just started your exam preparation today, when you opened this book . . . but, in reality, you have been getting ready for the AP test all year. The AP exam is an authentic test of your physics knowledge and skills. Your AP Physics *class* presumably is set up to teach those skills. So, don't give your class short shrift. Diligent attention to your class lectures, demonstrations, and assignments can only save you preparation time in the long run.

Of course, you may not be satisfied with the quantity or quality of your in-class instruction. And even if your class is the best in the country, you will still need a reminder of what you covered way back at the beginning of the year. That's where this book, and extracurricular AP exam preparation, are useful.

What Should I Study?

You will hear plenty of poorly-thought-out advice about how to deal with the vast amounts of material on the B exam. Fact is, in the month or two before the exam, you do not have enough time to re-teach yourself the entire course. So, you ask a presumed expert, "What should I study?"

Bad Answer Number 1: *"Everything."*

This logic says, every topic listed in the AP course description is guaranteed to show up somewhere on the exam, whether in the free-response or the multiple-choice sections. So, you must study everything. That's ridiculous, I say to my students. You've been studying "everything" all year. You need to focus your last-month study on those topics that are most likely to increase your score.

Bad Answer Number 2: *"Let me use my crystal ball to tell you exactly what types of problems will show up on this year's free-response exam. Study these."*

I know teachers who think they're oracles . . . "A *PV* diagram was on last year's test, so it won't be on this year's. And, we haven't seen point charges for two straight years, so we'll definitely see one this year."[1] Suffice it to say that a teacher who is not on the test development committee has no possible way of divining which specific types of problems will appear on the exam, any more than a college basketball "expert" can say with confidence which teams will make the final four. And, even if you *did* know which topics would be covered on the free-response section, all of the other topics must appear on the multiple-choice section! So don't choose your study strategy based on an oracle's word.

Good Answer: *Do a Cost–Benefit Analysis*

You know how much time you have left. Use that limited time to study the topics that are most likely to increase your score. The trick is identifying those topics. Start with honest, hyperbole-free answers to two questions, in the following manner.

Imagine that the AP Physics Genie[2] has granted you two boons. You may choose one type of problem that *will* be tested on the free-response exam; and you may choose one type of problem that will *not* appear on the free response. Now, answer:

1. What topic or problem type do you ask the genie to put on the exam?
2. What topic or problem type do you forbid the genie to put on the exam?

If you are extremely comfortable, say, solving kinematics and projectile problems, why would you spend any time on those? It won't hurt to give yourself a quick reminder of fundamental concepts, but in-depth study of what you know well is a waste of valuable time. On the other hand, if you're *un*comfortable with, say, buoyant forces, then spend a couple of evenings learning how to deal with them. Study the topics you're afraid will appear; relax about those you're best at.[3]

This is an important point—don't study so broadly that you don't get good at some specific type of problem. Use Chapter 8's drill exercises, or the end-of-chapter examples in this book, or some similar handout from your teacher, or a subset of your textbook's end-of-chapter problems, to keep practicing until you actually are *hoping* to see certain types of problems on your test. That's far more useful than just skimming around.

---

[1]A moment's thought will find some inconsistency in the above logic.

[2] . . . who is not a real person . . .

[3]I know many wise guys will say, "There's nothing I'm comfortable with; I'm bad at everything." That's called defeatism, and you shouldn't tolerate that from yourself. If you were to tell your softball coach, "Hey, I'm going to strike out at the plate, let grounders go through my legs, and drop all the fly balls hit to me," would the coach let you play? More likely, he or she would kick you off the team! When you pretend that you can't do anything in physics, you do yourself a tremendous disservice. Pick *something* that you can figure out, *some* topic you can develop confidence in, and go from there.

No Calculus, Remember

Physics B is a very broad course. But, there is *no calculus*; there's no need even to look at a problem that involves a derivative or an integral, even if your teacher or your textbook used some calculus. In fact, you shouldn't have to do any math more difficult than solving two simultaneous equations, or recognizing the definitions of the basic trig functions. Have an idea of the general level of math expected on AP problems. Then, if you're starting to do more complicated math than ever before, you know you're probably taking an incorrect approach—you're missing something simpler.

Have a Plan for the Exam

When it comes to the last few days before the exam, think about your mental approach. You can do very well on the exam even if you have difficulty with a few of the topics. But, know ahead of time which topics you are weak on. If you have trouble, say, with electric fields, plan on skipping electric fields multiple-choice questions so as to concentrate on those that you'll have more success on. Don't fret about this decision—just make it ahead of time, and follow your plan. On the free-response test, though, be sure to approach every problem. Sure, it's okay to decide that you will not waste time on electric fields. But if you read the entire problem, you might find that parts (d) and (e) are simple mechanics questions, or ask about some aspect of electricity that you understand just fine.

> **Exam tip from an AP Physics veteran:**
> I knew going into my exam that I couldn't handle double slit problems. Sure enough, problem 4 was about sound coming through two speakers and interfering—I knew this was like a double slit problem, so I was going to skip it. However, I looked at part (a) . . . it just asked for a wavelength using $v = \lambda f$! Well, I could do that just fine . . . and I could make an educated guess at the last three parts. I got a good bit of partial credit that probably helped me toward the 5 that I earned.
>
> —*Mark, college junior*

Three Different Study Schedules

MONTH	PLAN A: FULL SCHOOL YEAR	PLAN B: ONE SEMESTER	PLAN C: 6 WEEKS
September–October	Chapter 1–5	——	——
November	Chapters 9–12, Chapter 6	——	——
December	Chapters 13–15	——	——
January	Chapters 16–17	Chapters 1–5, Chapters 9–15, Chapter 6	——
February	Chapters 18–20	Chapters 1–5, Chapters 9–15	——
March	Chapters 21–22	Chapters 16–19	——
April	Chapters 23–24	Chapters 20–24	Skim Chapters 1–5, Rapid Review of Chapters 9–18, Skim Chapters 9–22, Chapter 6
May	Chapters 7–8; Review everything; Practice Exams	Chapters 7–8; Review everything; Practice Exams	Skim Chapters 23–24, Chapters 7–8; Review everything; Practice Exams

Calendar for Each Plan

Plan A: You Have a Full School Year to Prepare

Although its primary purpose is to prepare you for the AP Physics exam you will take in May, this book can enrich your study of physics, your analytical skills, and your problem-solving abilities.

SEPTEMBER–OCTOBER (Check off the activities as you complete them.)

☐ Determine the study mode (A, B, or C) that applies to you.

☐ Carefully read Steps 1 and 2 of this book.

☐ Work through the diagnostic exam.

☐ Get on the Web and take a look at the AP Web site(s).

☐ Skim Step 4. (Reviewing the topics covered in this section will be part of your yearlong preparation.)

☐ Buy a few color highlighters.

☐ Flip through the entire book. Break the book in. Write in it. Highlight it.

☐ Get a clear picture of what your own school's AP Physics curriculum is.

☐ Begin to use this book as a resource to supplement the classroom learning.

NOVEMBER (The first 10 weeks have elapsed.)

☐ Read and study Chapter 9, A Bit About Vectors.

☐ Read and study Chapter 10, Free-Body Diagrams and Equilibrium.

☐ Read and study Chapter 11, Kinematics.

☐ Read and study Chapter 12, Newton's Second Law, $F_{net} = ma$.

☐ Read Chapter 6, Memorizing Equations in the Shower.

DECEMBER

☐ Read and study Chapter 13, Momentum.

☐ Read and study Chapter 14, Energy Conservation.

☐ Read and study Chapter 15, Gravitation and Circular Motion.

☐ Review Chapters 9–12.

JANUARY (20 weeks have elapsed.)

☐ Read and study Chapter 16, Simple Harmonic Motion.

☐ Read and study Chapter 17, Thermodynamics.

☐ Review Chapters 9–15.

FEBRUARY

☐ Read and study Chapter 18, Fluid Mechanics.

☐ Read and study Chapter 19, Electrostatics.

☐ Read and study Chapter 20, Circuits.

☐ Review Chapters 9–17.

MARCH (30 weeks have now elapsed.)

☐ Read and study Chapter 21, Magnetism.

☐ Read and study Chapter 22, Waves.

☐ Review Chapters 9–20.

APRIL

☐ Read and study Chapter 23, Optics.

☐ Read and study Chapter 24, Atomic and Nuclear Physics.

☐ Review Chapters 9–22.

MAY (first 2 weeks) (THIS IS IT!)

☐ Review Chapters 9–24—all the material!!!

☐ Read Chapters 7–8 carefully!

☐ Take the Practice Exams, and score yourself.

☐ Get a good night's sleep before the exam. Fall asleep knowing that you are well prepared.

GOOD LUCK ON THE TEST.

Plan B: You Have One Semester to Prepare

Working under the assumption that you've completed one semester of your physics course, the following calendar will use those skills you've been practicing to prepare you for the May exam.

JANUARY–FEBRUARY
- ☐ Carefully read Steps 1 and 2 of this book.
- ☐ Work through the diagnostic exam.
- ☐ Read and study Chapter 9, A Bit About Vectors.
- ☐ Read and study Chapter 10, Free-Body Diagrams and Equilibrium.
- ☐ Read and study Chapter 11, Kinematics.
- ☐ Read and study Chapter 12, Newton's Second Law, $F_{net} = ma$.
- ☐ Read and study Chapter 13, Momentum.
- ☐ Read and study Chapter 14, Energy Conservation.
- ☐ Read and study Chapter 15, Gravitation and Circular Motion.
- ☐ Read Chapter 6, Memorizing Equations in the Shower.

MARCH (10 weeks to go)
- ☐ Read and study Chapter 16, Simple Harmonic Motion.
- ☐ Review Chapters 9–12.
- ☐ Read and study Chapter 17, Thermodynamics.

- ☐ Read and study Chapter 18, Fluid Mechanics.
- ☐ Read and study Chapter 19, Electrostatics.
- ☐ Review Chapters 13–15.

APRIL
- ☐ Read and study Chapter 20, Circuits.
- ☐ Read and study Chapter 21, Magnetism.
- ☐ Review Chapters 9–15.
- ☐ Read and study Chapter 22, Waves.
- ☐ Read and study Chapter 23, Optics.
- ☐ Read and study Chapter 24, Atomic and Nuclear Physics.
- ☐ Review Chapters 16–19.

MAY (first 2 weeks) (THIS IS IT!)
- ☐ Review Chapters 9–24—all the material!!!
- ☐ Read Chapters 7–8 carefully!
- ☐ Take the Practice Exams and score yourself.
- ☐ Get a good night's sleep before the exam. Fall asleep knowing that you are well prepared.

GOOD LUCK ON THE TEST.

Plan C: You Have Six Weeks to Prepare

At this point, we assume that you have been building your physics knowledge base for more than six months. You will, therefore, use this book primarily as a specific guide to the AP Physics exam. Given the time constraints, now is not the time to try to expand your AP Physics knowledge. Rather, you should focus on and refine what you already do know.

APRIL 1–15
- ☐ Skim Steps 1 and 2 of this book.
- ☐ Skim Chapters 9–16.
- ☐ Skim and highlight the Glossary at the end of the book.
- ☐ Read Chapter 6, and work on memorizing equations.

APRIL 16–MAY 1
- ☐ Skim Chapters 17–22.
- ☐ Carefully go over the Rapid Review sections of Chapters 9–16.
- ☐ Continue to work on memorizing equations.

MAY (first 2 weeks) (THIS IS IT!)
- ☐ Skim Chapters 23–24.
- ☐ Carefully go over the Rapid Review sections of Chapters 9–24.
- ☐ Read Chapters 7–8.
- ☐ Take the Practice Exams and score yourself.
- ☐ Get a good night's sleep before the exam. Fall asleep knowing that you are well prepared.

GOOD LUCK ON THE TEST.

STEP **2**

Determine Your Test Readiness

CHAPTER 4

Fundamentals Quizzes

IN THIS CHAPTER

Summary: To test your readiness for the exam, take these short quizzes on four fundamental topics of AP Physics.

Key Ideas

✪ Find out what you know—and what you don't know—about mechanics.

✪ Find out what you know—and what you don't know—about thermodynamics and fluid mechanics.

✪ Find out what you know—and what you don't know—about electricity and magnetism.

✪ Find out what you know—and what you don't know—about waves, optics, and atomic and nuclear physics.

✪ Focus your exam preparation time *only* on the areas you don't already know well.

These short quizzes may be helpful if you're looking for some additional review of the most fundamental topics in AP Physics. If you can get all these right, you are READY for the exam!

The answers are printed at the end of this chapter.

Mechanics Quiz

1. What is the mass of a block with weight 100 N? ⌣ 10kg

2. Give the equations for two types of potential energy, identifying each.

$\frac{1}{2}kx^2$ ~ spring ⌣ mgh ~ gravitational

3. When an object of mass m is on an incline of angle θ, one must break the weight of the object into components parallel and perpendicular to the incline.
   (a) What is the component of the weight parallel to the incline? _$mg\sin\theta$_
   (b) What is the component of the weight perpendicular to the incline? _$mg\cos\theta$_

4. Write two expressions for work, including the definition of work and the work-energy principle. *[handwritten: $W = F\Delta x$ Force over a distance $U = \Delta KE$]*

5. Quickly identify as a vector or a scalar:

   __V__ acceleration   __V__ force   __V__ momentum
   __V__ velocity   __S__ speed   __V__ displacement
   __S__ work   __S__ mass   __S__ kinetic energy

6. Name at least four things that can NEVER go on a free-body diagram.

7. Write two expressions for impulse. What are the units of impulse? *[handwritten: $F\Delta t$]*

8. In what kind of collision is momentum conserved? In what kind of collision is kinetic energy conserved? *[handwritten: both]*

9. What is the mass of a block with weight W? *[handwritten: W/g]*

10. A ball is thrown straight up. At the peak of its flight, what is the ball's acceleration? Be sure to give both magnitude and direction. *[handwritten: $-9.8\,m/s^2$ ground]*

11. A mass experiences a force vector with components 30 N to the right, 40 N down. Explain how to determine the magnitude and direction (angle) of the force vector. *[handwritten: $37°$ horizontal $50N$]*

12. Write the definition of the coefficient of friction, μ. What are the units of μ? *[handwritten: $f = \mu F_N$ none]*

13. How do you find acceleration from a velocity-time graph? *[handwritten: Slope]*

14. How do you find displacement from a velocity-time graph? *[handwritten: area]*

15. How do you find velocity from a position-time graph? *[handwritten: Slope]*

16. An object has a positive acceleration. Explain *briefly* how to determine whether the object is speeding up, slowing down, or moving with constant speed. *[handwritten: depends on direction of velocity, of friction?]*

17. Given the velocity of an object, how do you tell which direction that object is moving? *[handwritten: vector, right is positive, left neg]*

18. When is the gravitational force on an object mg? When is the gravitational force Gm_1m_2/r^2? *[handwritten: between two objects   for one object   field]*

19. What is the direction of the net force on an object that moves in a circle at constant speed? *[handwritten: towards center]*

20. Under what conditions is the equation $x - x_0 = v_0 t + \frac{1}{2}at^2$ valid? Give a specific situation in which this equation might seem to be valid, but is NOT. *[handwritten: a needs to be constant   depends on negative   car deceleration   ans]*

Thermodynamics and Fluid Mechanics Quiz

1. What is the equation for linear thermal expansion? What are the units for the coefficient of linear expansion? *[handwritten: $\Delta L = \alpha L_0(T_f - T_i)$]*

2. How do you determine the internal energy of a gas given the temperature of the gas? Define all variables in your equation. *[handwritten: $\Delta U = \frac{3}{2}nR\Delta t$]*

[marginal handwritten notes: $\frac{mv^2}{r} = \frac{Gm_1m_2}{r^2}$; $F_c = F_g$]

3. How do you determine the rms speed of molecules given the temperature of a gas? Define all variables in your equation. $\Delta U = KE = \frac{1}{2}mv_{rms}^2$

4. State the equation for the first law of thermodynamics. What does each variable stand for? What are the units of each term? $\Delta U = Q + W$

5. Sketch two isotherms on the *PV* diagram below. Label which isotherm represents the higher temperature.

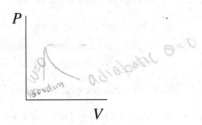

6. Describe a situation in which heat is added to a gas, but the temperature of the gas does *not* increase. $Q = nR\Delta T + W = \Delta U$ if expands

7. Imagine you are given a labeled *P-V* diagram for one mole of an ideal gas. *Note that one of the following is a trick question!*

  (a) How do you use the graph to determine how much work is done on or by the gas? $-P\Delta V$

  (b) How do you use the graph to determine the change in the gas's internal energy? $U = \frac{3}{2}nRT$

  (c) How do you use the graph to determine how much heat was added to or removed from the gas? can't

8. What is the definition of the efficiency of an ideal heat engine? How does the efficiency of a real engine relate to the ideal efficiency?

9. For the equation $P = P_0 + \rho gh$,

  (a) for what kind of situation is the equation valid? → underwater

  (b) what does P_0 stand for (careful!)? original pressure (atm)

10. Write Bernoulli's equation. $P_0 + \frac{1}{2}\rho v_1^2 + \rho gh = P_2 + \frac{1}{2}\rho v_2^2 + \rho gh_2$

11. State Archimedes' principle in words by finishing the following sentence: "The buoyant force on an object in a fluid is equal to . . ." w of water displaced

12. For a flowing fluid, what quantity does *Av* represent, and why is this quantity the same everywhere in a flowing fluid? volume flow rate

13. Write the alternate expression for mass which is useful when dealing with fluids of known density. $\rho V = m$

Electricity and Magnetism Quiz

  1. Given the charge of a particle and the electric field experienced by that particle, give the equation to determine the electric force acting on the particle.

  2. Given the charge of a particle and the magnetic field experienced by that particle, give the equation to determine the magnetic force acting on the particle.

3. What are the units of magnetic flux? What are the units of EMF?

4. A wire carries a current to the left, as shown below. What is the direction and magnitude of the magnetic field produced by the wire at point P?

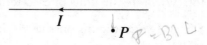

5. When is the equation kQ/r^2 valid? What is this an equation for?

6. The electric field at point P is 100 N/C; the field at point Q, 1 meter away from point P, is 200 N/C. A point charge of +1 C is placed at point P. What is the magnitude of the electric force experienced by this charge?

7. Can a current be induced in a wire if the flux through the wire is zero? Explain.

8. True or false: In a uniform electric field pointing to the right, a negatively charged particle will move to the left. If true, justify with an equation; if false, explain the flaw in reasoning.

9. Which is a vector and which is a scalar: electric field and electric potential?

10. Fill in the blank with either "parallel" or "series":

 (a) Voltage across resistors in _____ must be the same for each.
 (b) Current through resistors in _____ must be the same for each.
 (c) Voltage across capacitors in _____ must be the same for each.
 (d) Charge stored on capacitors in _____ must be the same for each.

11. A uniform electric field acts to the right. In which direction will each of these particles accelerate?

 (a) proton
 (b) positron (same mass as electron, but opposite charge)
 (c) neutron
 (d) anti-proton (same mass as proton, but opposite charge)

12. A uniform magnetic field acts to the right. In which direction will each of these particles accelerate, assuming they enter the field moving toward the top of the page?

 (a) proton
 (b) positron (same mass as electron, but opposite charge)
 (c) neutron
 (d) anti-proton (same mass as proton, but opposite charge)

13. How do you find the potential energy of an electric charge?

Waves, Optics, and Atomic and Nuclear Physics Quiz

1. (a) When light travels from water ($n = 1.3$) to glass ($n = 1.5$), which way does it bend?
 (b) When light travels from glass to water, which way does it bend?
 (c) In which of the above cases may total internal reflection occur?
 (d) Write (but don't solve) an equation for the critical angle for total internal reflection between water and glass.

2. In the equation

$$x = \frac{m\lambda L}{d}$$

describe in words what each variable means:

x _[handwritten] length form maximum – minima_
m _[handwritten] wave differential_
L _[handwritten] length between slits + board_
d _[handwritten] length between slits_
λ _[handwritten] wavelength_

3. The equation $d \sin \theta = m\lambda$ is also used for "light through a slit" types of experiments. When should this equation, rather than the equation in question 2, be used? _[handwritten] double slit   construct_

4. Describe two principal rays drawn for a convex lens. Be careful to distinguish between the *near* and *far* focal points.

1.
2.

5. Describe two principal rays drawn for a concave lens. Be careful to distinguish between the *near* and *far* focal points.

1.
2.

6. We often use two different equations for wavelength:

$$\lambda = \frac{hc}{\Delta E}, \text{ and } \lambda = \frac{h}{mv}$$

[handwritten] energy of wave   _[handwritten] speed of particle_

When is each used?

7. Name the only decay process that affects neither the atomic number nor the atomic mass of the nucleus. _[handwritten] β_

Answers to Mechanics Quiz

1. Weight is *mg*. So, mass is weight divided by *g*, which would be 100 N/(10 N/kg) = 10 kg.

2. PE = *mgh*, gravitational potential energy;
PE = ½ *kx²*, potential energy of a spring.

3. (a) *mg* sin θ is parallel to the incline.
(b) *mg* cos θ is perpendicular to the incline.

4. The definition of work is work = force times parallel displacement.
The work-energy principle states that net work = change in kinetic energy.

5. vectors: acceleration, force, momentum, velocity, displacement

scalars: speed, work, mass, kinetic energy

6. Only forces acting on an object and that have a single, specific source can go on free-body diagrams. Some of the things that cannot go on a free-body diagram but that students often put there by mistake:

motion	mass	acceleration	*ma*
centripetal force	velocity	inertia	

7. Impulse is force times time interval, and also change in momentum. Impulse has units either of newton·seconds or kilogram·meters/second.

8. Momentum is conserved in *all* collisions. Kinetic energy is conserved only in elastic collisions.

9. Using the reasoning from question #1, if weight is *mg*, then $m = W/g$.

10. The acceleration of a projectile is *always* g; i.e., 10 m/s², downward. Even though the velocity is instantaneously zero, the velocity is still changing, so the acceleration is *not* zero. (By the way, the answer "−10 m/s²" is wrong unless you have clearly and specifically defined the down direction as negative for this problem.)

11. The magnitude of the resultant force is found by placing the component vectors tip-to-tail. This gives a right triangle, so the magnitude is given by the Pythagorean theorem, 50 N. The angle of the resultant force is found by taking the inverse tangent of the vertical component over the horizontal component, $\tan^{-1}(40/30)$. This gives the angle measured from the horizontal.

12.
$$\mu = \frac{F_f}{F_n}$$

friction force divided by normal force. μ has no units.

13. Acceleration is the slope of a velocity-time graph.

14. Displacement is the area under a velocity-time graph (i.e., the area between the graph and the horizontal axis).

15. Velocity is the slope of a position-time graph. If the position-time graph is curved, then instantaneous velocity is the slope of the tangent line to the graph.

16. Because acceleration is not zero, the object *cannot* be moving with constant speed. If the signs of acceleration and velocity are the same (here, if velocity is positive), the object is speeding up. If the signs of acceleration and velocity are different (here, if velocity is negative), the object is slowing down.

17. An object *always* moves in the direction indicated by the velocity.

18. Near the surface of a planet, *mg* gives the gravitational force. Newton's law of gravitation, Gm_1m_2/r^2, is valid everywhere in the universe. (It turns out that *g* can be found by calculating GM_{planet}/R_{planet}^2, where R_{planet} is the planet's radius.)

19. An object in uniform circular motion experiences a *centripetal*, meaning "center seeking," force. This force must be directed to the center of the circle.

20. This and all three kinematics equations are valid only when acceleration is constant. So, for example, this equation can NOT be used to find the distance traveled by a mass attached to a spring. The spring force changes as the mass moves; thus, the acceleration of the mass is changing, and kinematics equations are not valid. (On a problem where kinematics equations aren't valid, conservation of energy usually is what you need.)

Answers to Thermodynamics and Fluid Mechanics Quiz

1. $\Delta L = \alpha L_0 \Delta T$. The units of α can be figured out by solving for

$$\alpha = \frac{\Delta L}{L_0 \Delta T}$$

The units of length cancel, and we're left with 1/K or 1/°C. (Either kelvins or degrees Celsius are acceptable here because only a change in temperature appears in the equation, not an absolute temperature.)

2. $U = \frac{3}{2} N k_B T$. Internal energy is $\frac{3}{2}$ times the number of molecules in the gas times Boltzmann's constant (which is on the constant sheet) times the absolute temperature, in kelvins. Or, $U = \frac{3}{2} n R T$ is correct, too, because $N k_B = n R$. (Capital N represents the number of molecules; small n represents the number of moles.)

3.
$$v_{rms} = \sqrt{\frac{3 k_B T}{m}}$$

k_B is Boltzmann's constant, T is absolute temperature in kelvins, and m is the mass of each molecule in kilograms (*NOT* in amu!). This can also be expressed as $\sqrt{\frac{3RT}{\mu}}$ where μ is the molar mass of the gas and R is the ideal gas constant.

4.
$$\Delta U = Q + W$$

Change in internal energy is equal to (say it in rhythm, now) "*heat added to, plus work done on*" a gas. Each term is a form of energy, so has units of joules.

5. The isotherm labeled as "2" is at the higher temperature because it's farther from the origin.

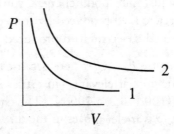

6. Let's put the initially room-temperature gas into a boiling water bath, adding heat. But let's also make the piston on the gas cylinder expand, so that the gas does work. By the first law of thermodynamics, if the gas does as much or more work than the heat added to it, then ΔU will be zero or negative, meaning the gas's temperature stayed the same or went down.

7. (a) Find the area under the graph. (b) Use $PV = nRT$ to find the temperature at each point; then, use $U = \frac{3}{2} nRT$ to find the internal energy at each point; then subtract to find ΔU. (c) You can NOT use the graph to determine heat added or removed. The only way to find Q is to find ΔU and W.

8. For an ideal heat engine,

$$e = \frac{T_h - T_c}{T_h}$$

A real heat engine will have a smaller efficiency than this.

9. (a) This is valid for a static (not moving) column of fluid.
   (b) P_0 stands for pressure at the top of the fluid; not necessarily, but sometimes, atmospheric pressure.

10. $P_1 + \rho g y_1 + \frac{1}{2}\rho v_1^2 = P_2 + \rho g y_2 + \frac{1}{2}\rho v_2^2$

11. . . . the weight of the fluid displaced.

12. Av is the volume flow rate. Fluid can't be created or destroyed; so, unless there's a source or a sink of fluid, total volume flowing past one point in a second must push the same amount of total volume past another downstream point in the same time interval.

13. mass = density · volume.

Answers to Electricity and Magnetism Quiz

1. $F = qE$.

2. $F = qvB \sin \theta$.

3. Magnetic flux is BA, so the units are tesla·meters^2 (or, alternatively, webers). Emf is a voltage, so the units are volts.

4. Point your right thumb in the direction of the current, i.e., to the left. Your fingers point in the direction of the magnetic field. This field wraps around the wire, pointing into the page above the wire and out of the page below the wire. Since point P is below the wire, the field points out of the page.

5. This equation is only valid when a point charge produces an electric field. (Careful— if you just said "point charge," you're not entirely correct. If a point charge experiences an electric field produced by something else, this equation is irrelevant.) It is an equation for the electric field produced by the point charge.

6. Do *not* use $E = kQ/r^2$ here because the electric field is known. So, the source of the electric field is irrelevant—just use $F = qE$ to find that the force on the charge is (1 C)(100 N/C) = 100 N. (The charge is placed at point P, so anything happening at point Q is irrelevant.)

7. Yes! Induced EMF depends on the *change* in flux. So, imagine that the flux is changing rapidly from one direction to the other. For a brief moment, flux will be zero; but flux is still changing at that moment. (And, of course, the induced current will be the EMF divided by the resistance of the wire.)

8. False. The negative particle will be *forced* to the left. But the particle could have entered the field while moving to the right . . . in that case, the particle would continue moving to the right, but would slow down.

9. Electric field is a vector, so fields produced in different directions can cancel. Electric potential is a scalar, so direction is irrelevant.

10. Voltage across resistors in parallel must be the same for each.
    Current through resistors in series must be the same for each.
    Voltage across capacitors in parallel must be the same for each.
    Charge stored on capacitors in series must be the same for each.

11. The positively charged proton will accelerate with the field, to the right.
    The positively charged positron will accelerate with the field, to the right.
    The uncharged neutron will not accelerate.
    The negatively charged anti-proton will accelerate against the field, to the left.

12. Use the right-hand rule for each:
    The positively charged proton will accelerate into the page.
    The positively charged positron will accelerate into the page.
    The uncharged neutron will not accelerate.
    The negatively charged anti-proton will accelerate out of the page.

13. If you know the electric potential experienced by the charge, $PE = qV$.

Answers to Waves, Optics, and Atomic and Nuclear Physics Quiz

1. (a) Light bends **toward** the normal when going from low to high index of refraction.
   (b) Light bends **away from** the normal when going from high to low index of refraction.
   (c) Total internal reflection can only occur when light goes from high to low index of refraction.
   (d) $\sin \theta_c = 1.3/1.5$

2. x is the distance from the central maximum to any other position, measured along the screen.
   m is the "order" of the point of constructive or destructive interference; it represents the number of extra wavelengths traveled by one of the interfering waves.
   L represents the distance from the double slit to the screen.
   d represents the distance between slits.
   λ represents the wavelength of the light.

3.
$$x = \frac{m\lambda L}{d}$$

   is used only when distance to the screen, L, is much greater than the distance between bright spots on the screen, x. $d\sin\theta = m\lambda$ can always be used for a diffraction grating or double-slit experiment, even if the angle at which you have to look for the bright spot is large.

4. For a convex (converging) lens:

   • The incident ray parallel to the principal axis refracts through the *far* focal point.
   • The incident ray through the *near* focal point refracts parallel to the principal axis.
   • The incident ray through the center of the lens is unbent.

   (Note that you don't necessarily need to know this third ray for ray diagrams, but it's legitimate.)

5. For a concave (diverging) lens:
   - The incident ray parallel to the principal axis refracts as if it came from the *near* focal point.
   - The incident ray toward the *far* focal point refracts parallel to the principal axis.
   - The incident ray through the center of the lens is unbent.

   (Note that you don't necessarily need to know this third ray for ray diagrams, but it's legitimate.)

6.
$$\lambda = \frac{hc}{\Delta E}$$

   is used to find the wavelength of a photon only. You can remember this because of the *c*, meaning the speed of light—only the massless photon can move at the speed of light.

$$\lambda = \frac{h}{mv}$$

   is the de Broglie wavelength of a massive particle. You can remember this because of the *m*—a photon has no mass, so this equation can never be used for a photon.

7. Gamma decay doesn't affect the atomic mass or atomic number. In gamma decay, a photon is emitted from the nucleus, but because the photon carries neither charge nor an atomic mass unit, the number of protons and neutrons remains the same.

What Do I Know, and What Don't I Know?

I'll bet you didn't get every question on all of the fundamentals quizzes correct. That's okay. The whole point of these quizzes is for you to determine where to focus your study.

It's a common mistake to "study" by doing 20 problems on a topic on which you are already comfortable. But that's not studying . . . that's a waste of time. You don't need to drill yourself on topics you already understand! It's also probably a mistake to attack what for you is the toughest concept in physics right before the exam. Virtually every student has that one chapter they just don't get, however hard they try. That's okay.

The fundamentals quizzes that you just took can tell you exactly what you should and should not study. Did you give correct answers with full confidence in the correctness of your response? In that case, you're done with that topic. No more work is necessary. The place to focus your efforts is on the topics where either you gave wrong answers that you thought were right, or right answers that you weren't really sure about.

Now, take the diagnostic test. Once you've used the fundamentals quizzes and diagnostic test to identify the specific content areas you want to work on, proceed to the review in Chapters 9–24. Read a chapter, work through the examples in the chapter, and attempt some of the problems at the end of the chapter. Then come back to these fundamentals quizzes. When you respond to every question confidently, you are ready.

CHAPTER 5

Take a Diagnostic Test

IN THIS CHAPTER

Summary: Assess your strengths and weaknesses by answering some sample questions and then reading the answers and explanations, so you'll know where to focus your efforts when preparing for the exam.

› Diagnostic Test

Kinematics

1. Which of the following must be true of an object that is slowing down?

   (A) Its acceleration must be negative.
   (B) Its velocity must be smaller than its acceleration.
   (C) It must experience more than one force.
   (D) Its acceleration and its velocity must be in opposite directions.
   (E) Its velocity must be negative.

2. A baseball is thrown straight up. It reaches a peak height of 15 m, measured from the ground, in a time of 1.7 s. Treating "up" as the positive direction, what is the acceleration of the ball when it reaches its peak height?

   (A) 0 m/s²
   (B) +8.8 m/s²
   (C) −8.8 m/s²
   (D) +9.8 m/s²
   (E) −9.8 m/s²

Newton's laws

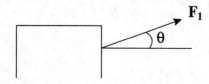

3. What is the vertical component of $\mathbf{F}_1$ in the above diagram?

   (A) ½$\mathbf{F}_1$
   (B) $\mathbf{F}_1$
   (C) $\mathbf{F}_1 \cos \theta$
   (D) $\mathbf{F}_1 \sin \theta$
   (E) $\mathbf{F}_1 \tan \theta$

4. The box pictured above moves at constant speed to the left. Which of the following is correct?

   (A) The situation is impossible. Because more forces act right, the block must move to the right.
   (B) $T_3 > T_1 + T_2$
   (C) $T_3 < T_1 + T_2$
   (D) $T_3 = T_1 + T_2$
   (E) A relationship among the three tensions cannot be determined from the information given.

The following diagram relates to Questions 5 and 6.

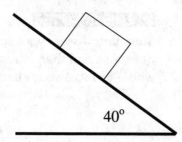

A block of mass m is sliding up a frictionless incline, as shown above. The block's initial velocity is 3 m/s up the plane.

5. What is the component of the weight parallel to the plane?

   (A) mg
   (B) $mg \cos 40°$
   (C) $mg \sin 40°$
   (D) $g \sin 40°$
   (E) $g \cos 40°$

6. What is the acceleration of the mass?

   (A) 3 m/s², up the plane
   (B) $mg \sin 40°$, up the plane
   (C) $mg \sin 40°$, down the plane
   (D) $g \sin 40°$, up the plane
   (E) $g \sin 40°$, down the plane

GO ON TO THE NEXT PAGE

Work/Energy

7. Which of the following is a scalar?

(A) velocity
(B) acceleration
(C) displacement
(D) kinetic energy
(E) force

8. A 500-g block on a flat tabletop slides 2.0 m to the right. If the coefficient of friction between the block and the table is 0.1, how much work is done on the block by the table?

(A) 0.5 J
(B) 1.0 J
(C) 0 J
(D) 100 J
(E) 50 J

9. A block has 1500 J of potential energy and 700 J of kinetic energy. Ten seconds later, the block has 100 J of potential energy and 900 J of kinetic energy. Friction is the only external force acting on the block. How much work was done on this block by friction?

(A) 600 J
(B) 200 J
(C) 1400 J
(D) 1200 J
(E) 120 J

Momentum

10. Two identical small balls are moving with the same speed toward a brick wall. After colliding with the wall, ball 1 sticks to the wall while ball 2 bounces off the wall, moving with almost the same speed that it had initially. Which ball experiences greater impulse?

(A) ball 1
(B) ball 2
(C) Both experience the same impulse.
(D) The answer cannot be determined unless we know the time of collision.
(E) The answer cannot be determined unless we know the force each ball exerts on the wall.

11. Ball A moves to the right with a speed of 5.0 m/s; ball B moves to the left with speed 2.0 m/s. Both balls have mass 1.0 kg. What is the total momentum of the system consisting only of balls A and B?

(A) 7.0 N·s to the right
(B) 3.0 N·s to the right
(C) zero
(D) 7.0 N·s to the left
(E) 3.0 N·s to the left

12. Momentum of an isolated system always remains constant. However, in a collision between two balls, a ball's momentum might change from, say, +1 kg m/s to −1 kg m/s. How can this be correct?

(A) It is *not* correct. Momentum conservation means that the momentum of an object must remain the same.
(B) A force outside the two-ball system must have acted.
(C) Friction is responsible for the change in momentum.
(D) Although one ball's momentum changed, the momentum of *both* balls in total remained the same.
(E) Momentum is conserved because the magnitude of the ball's momentum remained the same.

Circular motion

13. Which of the following must be true of an object in uniform circular motion?

(A) Its velocity must be constant.
(B) Its acceleration and its velocity must be in opposite directions.
(C) Its acceleration and its velocity must be perpendicular to each other.
(D) It must experience a force away from the center of the circle.
(E) Its acceleration must be negative.

Harmonic motion

14. A mass on a spring has a frequency of 2.5 Hz and an amplitude of 0.05 m. What is the period of the oscillations?

(A) 0.4 s
(B) 0.2 s
(C) 8 s
(D) 20 s
(E) 50 s

GO ON TO THE NEXT PAGE

15. A mass *m* oscillates on a horizontal spring of constant *k* with no damping. The amplitude of the oscillation is *A*. What is the potential energy of the mass at its maximum displacement?

(A) zero
(B) *mgh*
(C) *kA*
(D) $\frac{1}{2}mv^2$
(E) $\frac{1}{2}kA^2$

Gravitation

16. A satellite orbits the moon far from its surface in a circle of radius *r*. If a second satellite has a greater speed, yet still needs to maintain a circular orbit around the moon, how should the second satellite orbit?

(A) with a radius *r*
(B) with a radius greater than *r*
(C) with a radius less than *r*
(D) Only an eccentric elliptical orbit can be maintained with a larger speed.
(E) No orbit at all can be maintained with a larger speed.

Electrostatics

17. Which of the following statements about electric potential is correct?

(A) A proton experiences a force from a region of low potential to a region of high potential.
(B) The potential of a negatively charged conductor must be negative.
(C) If the electric field is zero at point *P*, then the electric potential at *P* must also be zero.
(D) If the electric potential is zero at point *P*, then the electric field at *P* must also be zero.
(E) The electric potential with respect to earth ground can be less than zero at all points on an isolated wire conductor.

Questions 18 and 19 refer to the diagram below.

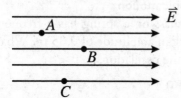

A uniform electric field points to the right, as shown above. A test charge can be placed at one of three points as shown in the above diagram.

18. At which point does the test charge experience the greatest force?

(A) point *A*
(B) point *B*
(C) point *C*
(D) The charge experiences the greatest force at two of these three points.
(E) The charge experiences the same force at all three points.

19. Which of the following is the correct ranking of the electric potential at each point, from greatest to least?

(A) ABC
(B) ACB
(C) CBA
(D) BCA
(E) A = B = C

20. An electron in an electric field is suspended above the earth's surface. Which of the following diagrams correctly shows the forces acting on this electron?

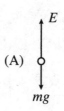

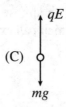

Circuits

21. Which of the following will increase the capacitance of a parallel plate capacitor?

(A) increasing the charge stored on the plates
(B) decreasing the charge stored on the plates
(C) increasing the separation between the plates
(D) decreasing the separation between the plates
(E) decreasing the area of the plates

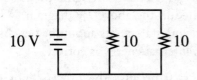

22. A 10-V battery is connected to two parallel 10-Ω resistors, as shown above. What is the current through and voltage across each resistor?

	Current	Voltage
(A)	1 A	5 V
(B)	1 A	10 V
(C)	0.5 A	5 V
(D)	2 A	10 V
(E)	2 A	5 V

Magnetic fields and force

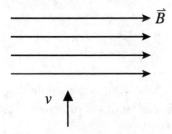

23. A positive point charge enters a uniform rightward magnetic field with a velocity v, as diagramed above. What is the direction of the magnetic force on the charge?

(A) in the same direction as v
(B) to the right
(C) to the left
(D) out of the page
(E) into the page

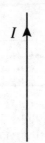

24. A long wire carries a current I toward the top of the page. What is the direction of the magnetic field produced by this wire to the left of the wire?

(A) into the page
(B) out of the page
(C) toward the bottom of the page
(D) toward the top of the page
(E) to the right

Electromagnetism

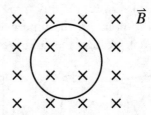

25. A circular loop of wire in the plane of the page is placed in a magnetic field pointing into the page, as shown above. Which of the following will NOT induce a current in the loop?

(A) moving the wire to the right in the plane of the page
(B) increasing the area of the loop
(C) increasing the strength of the magnetic field
(D) rotating the wire about a diameter
(E) turning the magnetic field off

GO ON TO THE NEXT PAGE

Fluids

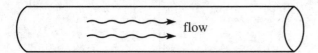

26. According to Bernoulli's principle, for a horizontal, uniform fluid flow in the thin, closed pipe shown above, which of the following statements is valid?

(A) The fluid pressure is greatest at the left end of the pipe.
(B) The fluid pressure is greatest at the right end of the pipe.
(C) The fluid pressure is the same at the left end of the pipe as it is at the right end of the pipe.
(D) At the point in the pipe where the fluid's speed is lowest, the fluid pressure is greatest.
(E) At the point in the pipe where the fluid's speed is greatest, the fluid pressure is lowest.

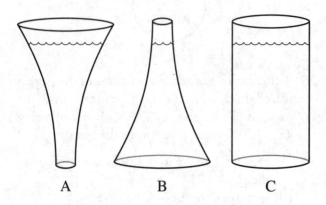

A   B   C

27. The three containers shown above contain the same depth of oil. Which container experiences the greatest pressure at the bottom?

(A) All experience the same pressure at the bottom.
(B) The answer depends on which has the greater maximum diameter.
(C) container A
(D) container B
(E) container C

Thermodynamics

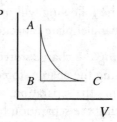

28. An ideal gas is taken through a process ABC as represented in the above PV diagram. Which of the following statements about the net change in the gas's internal energy is correct?

(A) $\Delta U = 0$ because the process ends on the same isotherm it started on.
(B) $\Delta U = 0$ because no heat was added to the gas.
(C) $\Delta U < 0$ because net work was done on the gas.
(D) $\Delta U > 0$ because net work was done by the gas.
(E) $\Delta U = 0$ because no work was done by the gas.

29. A heat engine is 20% efficient. If the engine does 500 J of work every second, how much heat does the engine exhaust every second?

(A) 2000 J
(B) 2500 J
(C) 100 J
(D) 400 J
(E) 500 J

Waves

30. Which of the following is a possible diagram of standing waves in an organ pipe that is closed at only one end?

(A)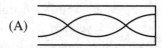

(B)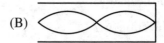

(C)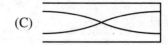

(D)

(E)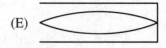

GO ON TO THE NEXT PAGE

31. An earthquake produces both transverse waves and longitudinal waves, both of which travel in a straight line to a seismograph. The waves are detected by the seismograph 500 km away from where the waves were created. Which of the following quantities must be the same for both the transverse and the longitudinal waves?

(A) their wavelength
(B) their amplitude
(C) the distance they traveled to the seismograph
(D) the time they took to travel to the seismograph
(E) their speed

Optics

32. Which of these rays can NOT be drawn for a concave spherical mirror?

(A) a ray through the mirror's center point reflecting straight back
(B) a ray through the focal point reflecting parallel to the principal axis
(C) a ray parallel to the principal axis reflecting through the focal point
(D) a ray through the mirror's center point reflecting parallel to the principal axis
(E) a ray to the intersection of the principal axis and the mirror itself, reflecting at an equal angle from the axis

33. The critical angle for the interface between water ($n = 1.3$) and glass ($n = 1.5$) is 60 degrees. What does this mean?

(A) Light cannot pass from water to glass unless the angle of incidence is less than 60°.
(B) Light cannot pass from glass to water unless the angle of incidence is less than 60°.
(C) Light cannot pass from water to glass unless the angle of incidence is more than 60°.
(D) Light cannot pass from glass to water unless the angle of incidence is more than 60°.
(E) Light cannot pass from glass to water at all.

Atomic and nuclear physics

34. When a nucleus undergoes alpha decay, its mass

(A) does not change
(B) decreases by 2 amu
(C) decreases by 4 amu
(D) increases by 2 amu
(E) increases by 4 amu

35. An electron in an atom absorbs a 620-nm photon. If the electron started with energy $E_o = -3.0$ eV, what was the electron's energy after the absorption?

(A) +2.0 eV
(B) +1.0 eV
(C) −1.0 eV
(D) −2.0 eV
(E) −5.0 eV

END OF DIAGNOSTIC TEST

› Answers and Explanations

1. D—Choices A and E don't make sense because the direction of an object's acceleration or velocity is essentially arbitrary—when solving a problem, you can usually pick the "positive" direction based on convenience. So neither value *must* be negative. We can rule out choice B because we know that a fast moving object can slow down very gradually. And there's no reason why you need multiple forces to make an object slow down, so that gets rid of choice C.

2. E—When an object is thrown in the absence of air resistance near the surface of the Earth, its acceleration in the vertical direction is always g, the acceleration due to gravity, which has a magnitude of 9.8 m/s². The acceleration due to gravity is directed down, toward the Earth. So the ball's acceleration is −9.8 m/s².

3. D—The vertical component of a vector is the magnitude of the vector times the sine of the angle measured to the horizontal; in this case, $F_1 \sin \theta$.

4. D—Something that moves in a straight line at constant speed is in equilibrium. So, the sum of left forces has to equal the sum of right forces.

5. C—On an incline, the weight vector parallel to the plane goes with the sine of the plane's angle. $g \sin 40°$ is an acceleration, not a weight.

6. E—Because the weight is the only force acting parallel to the plane, $mg \sin 40° = ma$, so $a = g \sin 40°$. This acceleration is down the plane, in the direction of the net force. Yes, the block is moving up the plane, but the block is slowing down, and so the acceleration must be in the opposite direction from the velocity.

7. D—All forms of energy are scalar quantities: potential energy, kinetic energy, work, internal energy of a gas. Energy doesn't have direction.

8. B—Work is force times parallel displacement. The force acting here is the force of friction, and the displacement is 2.0 m parallel to the force of friction. The friction force is equal to the coefficient of friction (0.10) times the normal force. The normal force in this case is equal to the block's weight of 5 N (because no other vertical forces act). Combining all these equations, the work done is (2.0 m) (0.1)(5 N) = 1.0 J.

9. D—Look at the total energy of the block, which is equal to the potential energy plus the kinetic energy. Initially, the total energy was 2200 J. At the end, the total energy was 1000 J. What happened to the extra 1200 J? Because friction was the only external force acting, friction must have done 1200 J of work.

10. B—Impulse is equal to the change in an object's momentum. Ball 1 changes its momentum from something to zero. But ball 2 changes its momentum from something to zero, and *then* to something in the other direction. That's a bigger momentum change than if the ball had just stopped. (If we had been asked to find the force on the ball, then we'd need the time of collision, but the impulse can be found without reference to force or time.)

11. B—Momentum is a *vector*, meaning direction is important. Call the rightward direction positive. Ball A has +5 kg·m/s of momentum; ball B has −2 kg·m/s of momentum. Adding these together, we get a total of +3 kg·m/s. This answer is equivalent to 3 N·s to the right. (The units kg·m/s and N·s are identical.)

12. D—The law of conservation of momentum requires that *all* objects involved in a collision be taken into account. An object can lose momentum, as long as that momentum is picked up by some other object.

13. C—"Uniform" circular motion means that an object's *speed* must be constant. But *velocity* includes direction, and the direction of travel changes continually. The velocity of the object is always along the circle, but the acceleration is centripetal; i.e., center-seeking. The direction toward the center of the circle is perpendicular to the path of the object everywhere.

14. A—Period is equal to 1/frequency, regardless of the amplitude of harmonic motion.

15. E—The maximum displacement is the amplitude. Energy of a spring is $\frac{1}{2} kx^2$. So, at $x = A$, the energy is $\frac{1}{2} kA^2$.

16. C—In an orbit, gravity provides a centripetal force. So, $GmM/r^2 = mv^2/r$. Solving for v,

$$v = \sqrt{\frac{GM}{r}}$$

where M is the mass of the moon. If the speed gets bigger, then the radius of orbit (in the denominator) must get smaller to compensate.

17. E—Most of these statements drive at the fundamental principle that the value of an electric potential can be set to anything; it is only the *difference* in electric potential between two points that has a physical usefulness. Usually potential is set to zero either at the ground or, for isolated point charges, a very long distance away from the charges. But potential can, in fact, be set to zero anywhere, meaning that the potential could easily be less than zero everywhere on a wire. (And a proton, a positive charge, is forced from high to low potential; not the other way around.)

18. E—This is a *uniform* electric field. The force on a charge in an electric field is given by $F = qE$. Therefore, as long as the electric field is the same at all three points, the force on the charge is the same as well.

19. B—Electric fields point from high to low potential. Since the field points to the right, rank the potentials from left to right—that is, ACB.

20. C—Only forces can go on free-body diagrams, and the electric field is not itself a force. The force *provided by* an electric field is qE; the weight of the electron is mg.

21. D—Capacitance is a property of the structure of the capacitor. Changing the charge on (or the voltage across) a capacitor does not change the capacitance. The capacitance of a parallel-plate capacitor is given by the equation

$$C = \frac{\varepsilon_0 A}{d}$$

Decreasing d, the distance between the plates, will increase the capacitance.

22. B—These resistors are in parallel with the battery; thus, they *both* must take the voltage of the battery, 10 V. The total current in the circuit is 2.0 A, but that current splits between the two resistors, leaving 1.0 A through each. This can also be determined by a direct application of Ohm's law—because we know both the voltage and resistance for each resistor, divide V/R to get the current.

23. E—Use the right-hand rule for the force on a charged particle in a magnetic field: point your right hand in the direction of the velocity, curl your fingers toward the magnetic field, and your thumb points into the page. The charge is positive, so your thumb points in the direction of the force.

24. B—This question uses the right-hand rule for the magnetic field produced by a current-carrying wire. Point your thumb in the direction of the current; your fingers wrap around the wire in the direction of the magnetic field. To the wire's left, your fingers point out of the page.

25. A—Only a changing magnetic flux induces a current. Flux is given by $BA\cos\theta$, where B is the magnetic field, and A is the area of the loop of wire. Obviously, then, choices B, C, and E change the flux and induce a current. Choice D produces a flux by changing θ, the angle at which the field penetrates the loop of wire. In choice A, no current is induced because the field doesn't change and always points straight through the loop.

26. C—One of the important results of Bernoulli's equation is that where a fluid flows faster, the fluid's pressure is lower. But here the fluid is flowing uniformly, so the pressure is the same everywhere.

27. A—The pressure caused by a static column of fluid is $P = P_0 + \rho gh$. The shape or width of a container is irrelevant—the pressure at the bottom depends only on the depth of the fluid. All of these fluids have equal depth, and so have equal pressures at the bottom.

28. A—Anytime a thermodynamic process starts and ends on the same isotherm, the change in internal energy is zero. In this case, the process starts and ends at the same point on the diagram, which fulfills the criterion of being on the same isotherm. (Net work was done on the gas because there is area within the cycle on the PV diagram, but that doesn't tell us anything about the change in internal energy.)

29. A—Efficiency is the useful work output divided by the heat input to an engine. So, in one second, 500 J of useful work was created. At 20% efficiency, this means that every second, 2500 J of heat had to be input to the engine. Any heat not used to do work is exhausted to the environment. Thus, 2000 J are exhausted.

30. D—A wave in a pipe that is closed at one end must have a node at one end and an antinode at the other. Choices A and C have antinodes at both ends; whereas, choices B and E have nodes at both ends.

31. C—Although the material through which waves travel determines the wave speed, longitudinal and transverse waves can have different speeds in the same medium. So the waves can take different times to go 500 km. Nothing stated gives any indication of a wavelength or amplitude (500 km is a distance traveled, not a wavelength), but both waves were produced at the same spot, and have to pass the seismograph; so both waves traveled the same distance.

32. D—Rays through the center point of a spherical mirror always reflect right back through the center point. (Note that the ray described in choice E is a legitimate principal ray, but not one that we discuss in the text nor that is necessary to remember.)

33. B—Total internal reflection can only occur when a light ray tries to travel from high to low index of refraction, so only B or D could be correct. Total internal reflection occurs when the angle of incidence (measured from the normal) is greater than the critical angle.

34. C—An amu is an atomic mass unit, meaning the mass of a proton or neutron. An alpha particle has 2 neutrons and 2 protons, so has 4 amu. When a nucleus undergoes alpha decay, it emits an alpha particle, and so loses 4 amu.

35. C—The energy of a photon is hc/λ. Especially without a calculator, it is necessary to use $hc = 1240$ eV·nm (this value is on the constant sheet). So, the energy absorbed by this electron is $1{,}240/620 = 2.0$ eV. The electron gained this much energy, so its energy level went from -3.0 eV to -1.0 eV.

Interpretation: How Ready Are You?

Now that you have finished the diagnostic exam and checked your answers, it is time to try to figure out what it all means. First, remember that getting only 60–65% of the answers correct will lead you to a 5 on the AP exam; about 30–40% correct is the criterion for a qualifying score of 3. You're not supposed to get 90% correct! So relax and evaluate your performance dispassionately.

Next, see if there are any particular areas in which you struggled. For example, were there any questions that caused you to think something such as, "I learned this . . . *when?!?*" *or* "What the heck is this?!?" If so, put a little star next to the chapter that contains the material in which this occurred. You may want to spend a bit more time on that chapter during your review for this exam. It is quite possible that you *never* learned some of the material in this book. Not every class is able to cover all the same information.

In general, try to interpret your performance on this test in a productive manner. If you did well, that's terrific . . . but don't get overconfident now. There's still a lot of material to review before you take the Practice Exams in Step 5—let alone the real AP exam. If you don't feel good about your performance, now is the time to turn things around. You have a great opportunity here—time to prepare for the real exam, a helpful review book, and a sense of what topics you need to work on most—so use it to its fullest. Good luck!

STEP **3**

Develop Strategies
for Success

CHAPTER 6

Memorizing Equations in the Shower

IN THIS CHAPTER

Summary: Learn how to memorize all the equations you absolutely need to know to ace the AP Physics exam.

Key Ideas
✪ Learn why memorizing equations is so critical.
✪ Learn equations by using them: practice solving problems without looking up the equations you need.
✪ Use mnemonic devices to help you remember.
✪ Practice speed: see how many equations you can say in four minutes.
✪ Use visual reminders: put a copy of the equation sheet somewhere you'll see it often.

Can You Ace This Quiz?

<u>Instructions:</u> We'll give you a prompt, you tell us the equation. Once you've finished, check your answers with the key at the end of this chapter.

1. Coefficient of friction in terms of F_f
2. Momentum
3. An equation for impulse
4. Another equation for impulse
5. An equation for mechanical power
6. Another equation for mechanical power
7. Lensmaker's equation
8. Ohm's law
9. An equation for work
10. Another equation for work
11. Snell's law
12. Power in a circuit
13. Period of a mass on a spring
14. F_{net}
15. Three kinematics equations
16. rms speed
17. Centripetal acceleration
18. Kinetic energy
19. Gravitational force
20. Magnetic field around a long, straight, current-carrying wire

So, How Did You Do?

Grade yourself according to this scale.

20 right	Excellent
0–19 right	Start studying

You may think we're joking about our grading system, but we're completely serious. Knowing your equations is absolutely imperative. Even if you missed one question on the quiz, you need to study your equations. Right now! A student who is ready to take the AP exam is one who can ace an "equations quiz" without even thinking about it. How ready are you?

Equations Are Crucial

It's easy to make an argument against memorizing equations. For starters, you're given all the equations you need when taking the free-response portion of the AP exam. And besides, you can miss a whole bunch of questions on the test and still get a 5.

But equations are the nuts and bolts of physics. They're the fundamentals. They should be the foundation on which your understanding of physics is built. Not knowing an equation—even one—shows that your knowledge of physics is incomplete. And every question on the AP exam assumes *complete* knowledge of physics.

Remember, also, that you don't get an equation sheet for the multiple-choice portion of the test. Put simply, you won't score well on the multiple-choice questions if you haven't memorized all of the equations.

What About the Free-Response Section?

The free-response questions test your ability to solve complex, multistep problems. They also test your understanding of equations. You need to figure out which equations to use when and how. The makers of the test are being nice by giving you the equation sheet— they're reminding you of all the equations you already know in case you cannot think of that certain equation that you *know* would be just perfect to solve a certain problem. But the sheet is intended to be nothing more than a reminder. It will not tell you when to use an equation or which equation would be best in solving a particular problem. You have to know that. And you will know that only if you have intimate knowledge of every equation.

Exam tip from an AP Physics veteran:
Don't use the equation sheet to "hunt and peck." The sheet can remind you of subtle things; for example, does the magnetic field due to a wire have an r or an r^2 in the denominator? But if you don't have the general idea that the magnetic field depends on current and gets weaker farther away from a wire, then you won't recognize

$$B = \frac{\mu_0 I}{2\pi r}$$

even if you go hunting for it.

—Wyatt, college freshman in engineering

Some Examples

We mentioned in Step 2 that some questions on the AP exam are designed solely to test your knowledge of equations. If you know your equations, you will get the question right. Here's an example.

A pendulum of length L swings with a period of 3 s. If the pendulum's length is increased to $2L$, what will its new period be?

(A) $3/\sqrt{2}$ s

(B) 3 s

(C) $3\sqrt{2}$ s

(D) 6 s

(E) 12 s

The answer is (C). The equation for a pendulum's period is

$$T = 2\pi\sqrt{\frac{L}{g}}$$

so if L increases by 2, the period must increase by $\sqrt{2}$.

Of course, the multiple-choice section will not be the only part of the exam that tests your knowledge of equations. Often, a part of a free-response question will also test your ability to use an equation. For example, check out this problem.

One mole of gas occupies 5 m³ at 273 K.

(a) At what pressure is this gas contained?

Yes, later in the problem you find one of those PV diagrams you hate so much. However, you can still score some easy points here if you simply remember that old standby, $PV = nRT$. Memorizing equations will earn you points. It's that simple.

Treat Equations like Vocabulary

Think about how you would memorize a vocabulary word: for example, "boondoggle." There are several ways to memorize this word. The first way is to say the word out loud and then spell it: "Boondoggle: B-O-O-N-D-O-G-G-L-E." The second way is to say the word and then say its definition: "Boondoggle: An unproductive or impractical project, often involving graft." If you were to use the first method of memorizing our word, you would become a great speller, but you would have no clue what "boondoggle" means. As long as you are not preparing for a spelling bee, it seems that the second method is the better one.

This judgment may appear obvious. Who ever learned vocabulary by spelling it? The fact is, this is the method most people use when studying equations.

Let's take a simple equation, $v_f = v_o + at$. An average physics student will memorize this equation by saying it aloud several times, and that's it. All this student has done is "spelled" the equation.

But you're not average.[1] Instead, you look at the equation as a whole, pronouncing it like a sentence: "*Vf* equals *v* naught plus *at*." You then memorize what it means and when to use it: "This equation relates initial velocity with final velocity. It is valid only when acceleration is constant." If you are really motivated, you will also try to develop some intuitive sense of why the equation works. "Of course," you say, "this makes perfect sense! Acceleration is just the change in velocity divided by the change in time. If acceleration is multiplied by the change in time, then all that's left is the change in velocity

$$\frac{\Delta v}{\Delta t} \cdot \Delta t = \Delta v$$

So the final velocity of an object equals its initial velocity plus the change in velocity."

The first step in memorizing equations, then, is to learn them as if you were studying for a vocabulary test, and not as if you were studying for a spelling bee.

Helpful Tips

Memorizing equations takes a lot of time, so you cannot plan on studying your equations the night before the AP exam. If you want to really know your equations like the back of your hand, you will have to spend months practicing. But it's really not that bad. Here are four tips to help you out.

Tip 1: Learn through use. Practice solving homework problems without looking up equations.

Just as with vocabulary words, you will only learn physics equations if you use them on a regular basis. The more you use your equations, the more comfortable you will be with them, and that comfort level will help you on the AP test.

The reason you should try solving homework problems without looking up equations is that this will alert you to trouble spots. When you can look at an equations sheet, it's easy to fool yourself into a false sense of confidence: "Oh, yeah, I *knew* that spring potential energy is ½ kx^2." But when you don't have an equations sheet to look at, you realize that either you know an equation or you don't. So if you solve homework problems without looking up equations, you'll quickly figure out which ones you know and which you don't; and then you can focus your studying on those equations that need more review.

Tip 2: Use mnemonic devices.

Use whatever tricks necessary to learn an equation. For example, it is often hard to remember that the period of a pendulum is

$$T = 2\pi \sqrt{\frac{L}{g}}$$

---

[1]In fact, just because you bought this book, we think that you're way better than average. "Stupendous" comes to mind. "Extraordinary, Gullible." Er . . . uh . . . cross that third one out.

and *not*

$$T = 2\pi\sqrt{\frac{g}{L}}$$

So make up some trick, like "The terms go in backward alphabetical order: **Two**-pi **r**oot *L* over *g*." Be creative.

Tip 3: The 4-Minute Drill.

Practice speed. Say the equations as fast as you can, then say them faster. Start at the top of the AP equations sheet[2] and work your way down. Have someone quiz you. Let that person give you a lead, like "Period of a pendulum," and you respond "Two-pi root *L* over *g*." See how many equations you can rattle off in four minutes. We call it the 4-Minute Drill.

This is much more fun with a group; for example, try to persuade your teacher to lead the class in a 4-minute drill. Not only will you get out of four minutes of lecture, but you may also be able to bargain with your teacher: "Sir, if we can rattle off 50 equations in the 4-Minute Drill, will you exempt us from doing tonight's problems?"[3]

Tip 4: Put a copy of the equations sheet somewhere visible.

See how the equations sheet looks next to your bathroom mirror. Or in your shower (laminated, of course). Or taped to your door. Or hung from your ceiling. You'd be surprised how much sparkle it can add to your decor. You'd also be surprised how easy it will be to memorize equations if you are constantly looking at the equations sheet.

So what are you waiting for? Start memorizing!

---

[2]We've included a copy of this sheet at the end of the book, along with a sheet of prompts to guide you through a 4-minute drill.

[3]"No."

❯ Answer Key to Practice Quiz

1. $\mu = \dfrac{F_f}{F_n}$

2. $p = mv$
3. $I = \Delta p$
4. $I = F\Delta t$
5. $P = Fv$

6. $P = \dfrac{W}{t}$

7. $\dfrac{1}{d_i} + \dfrac{1}{d_o} = \dfrac{1}{f}$

8. $V = IR$
9. $W = Fd$
10. $W = \Delta KE$
11. $n_1 \sin\theta_1 = n_2 \sin\theta_2$
12. $P = IV$

13. $T = 2\pi\sqrt{\dfrac{m}{k}}$

14. $F_{net} = ma$

15. $v_f = v_o + at$
$x - x_0 = v_0 t + \frac{1}{2}at^2$
$v_f^2 = v_0^2 + 2a(x - x_0)$

16. $v_{rms} = \sqrt{\dfrac{3k_B T}{m}}$

17. $a_c = \dfrac{v^2}{r}$

18. $KE = \frac{1}{2}mv^2$

19. $F = \dfrac{Gm_1 m_2}{r^2}$

20. $B_{wire} = \dfrac{\mu_0 I}{2\pi r}$

CHAPTER 7

How to Approach Each Question Type

IN THIS CHAPTER

Summary: Become familiar with the three types of questions on the exam: multiple-choice, free-response, and lab questions. Pace yourself, and know when to skip a question.

Key Ideas
✪ You can't use a calculator to answer multiple-choice questions because you don't need a calculator to figure out the answer.
✪ There are five categories of multiple-choice questions. Two of these involve numbers: easy calculations and order-of-magnitude estimates. The other three don't involve numbers at all: proportional reasoning questions, concept questions, and questions asking for the direct solution with variables only.
✪ Free-response questions test your understanding of physics, not obscure theories or technical terms.
✪ You can get partial credit on free-response questions.
✪ Each free-response section will contain at least one question that involves experiment design and analysis—in other words, a lab question.
✪ Check out our six steps to answering lab questions successfully.

How to Approach the Multiple-Choice Section

You cannot use a calculator on the multiple-choice section of the AP exam. "No calculator!" you exclaim. "That's inhumane!" Truth be told, you are not alone in your misguided fears. Most students are totally distraught over not being able to use their calculators. They think it's entirely unfair.

But the fact is: the AP test is very, very fair. There are no trick questions on the AP exam, no unreasonably difficult problems, and no super-tough math. So when the test says, "You may not use a calculator on this section," what it really means is, "You DON'T NEED to use a calculator on this section." In other words, the writers of the test are trying to communicate two very important messages to you. First, no question on the multiple-choice section requires lots of number crunching. Second,

> Physics is NOT about numbers.

Yes, you must use numbers occasionally. Yet you must understand that the *number* you get in answer to a question is always subordinate to what that number represents.

Many misconceptions about physics start in math class. There, your teacher shows you how to do a type of problem, then you do several variations of that same problem for homework. The answer to one of these problems might be 30,000,000, another 16.5. It doesn't matter . . . in fact, the book (or your teacher) probably made up random numbers to go into the problem to begin with. The "problem" consists of manipulating these random numbers a certain way to get a certain answer.

In physics, though, *every number has meaning*. Your answer will not be 30,000,000; the answer may be 30,000,000 electron-volts, or 30,000,000 seconds, but not just 30,000,000. If you don't see the difference, you're missing the fundamental point of physics.

We use numbers to represent REAL goings on in nature. 30,000,000 eV (or, 30 MeV) is an energy; this could represent the energy of a particle in a multibillion dollar accelerator, but it's much too small to be the energy of a ball dropped off of a building. 30,000,000 seconds is a time; not a few hours or a few centuries, but about one year. These two "30,000,000" responses mean entirely different things. If you simply give a number as an answer, you're doing a math problem. It is only when you can explain the meaning of any result that you may truly claim to understand physics.

So How Do I Deal with All the Numbers on the Test?

You see, the test authors aren't dumb—they are completely aware that you have no calculator. Thus, a large majority of the multiple-choice questions involve *no numbers at all!* And those questions that do use numbers will never expect more than the simplest manipulations. Here is a question you will **never** see on the AP test:

What is the magnitude of the magnetic field a distance of 1.5 m away from a long, straight wire that carries 2.3 A of current?

(A) 3.066×10^{-6} T
(B) 3.166×10^{-6} T
(C) 3.102×10^{-6} T
(D) 2.995×10^{-6} T
(E) 3.109×10^{-6} T

Yes, we know you might have seen this type of problem in class. But it will *not* be on the AP exam. Why not? Plugging numbers into a calculator is not a skill being tested by this examination. (You should have recognized that the equation necessary to solve this problem is

$$B = \frac{\mu_0 I}{2\pi r}$$

though.) We hope you see that, without a calculator, it is pointless to try to get a precise numerical answer to this kind of question.

Fine . . . Then What Kinds of Question *Will* Be Asked on the Multiple-Choice Section?

Fair enough. We break down the kinds of questions into five categories. First, the categories of questions that involve numbers:

1. easy calculations
2. order of magnitude estimates

Most questions, though, do NOT involve numbers at all. These are:

3. proportional reasoning
4. concept questions, subdivided into
   a. "Why?" questions, and
   b. diagram questions
5. direct solution with variables

Okay, let's take a look at a sample of each of these.

Easy Calculations

These test your knowledge of formulas.

A ball is dropped from a 45-m-high platform. Neglecting air resistance, how much time will it take for this ball to hit the ground?

(A) 1.0 s
(B) 2.0 s
(C) 3.0 s
(D) 4.0 s
(E) 5.0 s

You should remember the kinematics equation: $x - x_0 = v_0 t + \frac{1}{2} a t^2$. Here the initial velocity is zero because the ball was "dropped." The distance involved is 45 meters, and the acceleration is caused by gravity, 10 m/s². The solution must be found without a calculator, but notice how easy they have made the numbers:

$$45 \text{ m} = 0t + \frac{1}{2}(10 \text{ m/s}^2) \cdot t^2$$

$$90 = 10t^2$$

$$9 = t^2$$

$$t = 3.0 \text{ s}$$

Everything here can be done easily without a calculator, especially if you remember to use 10 m/s² for g. No problem!

Order of Magnitude Estimates

These test your understanding of the size of things, measurements, or just numbers.

A typical classroom is approximately cubic, with a front-to-back distance of about 5 meters. Approximately how many air molecules are in a typical classroom?

(A) 10^8
(B) 10^{18}
(C) 10^{28}
(D) 10^{38}
(E) 10^{48}

Wow, at first you have no idea. But let's start by looking at the answer choices. Notice how widely the choices are separated. The first choice is 100 million molecules; the next choice is a billion billion molecules! **Clearly no kind of precise calculation is necessary here.**

We can think of several ways to approach the problem. The slightly longer way involves a calculation using the ideal gas law, $PV = Nk_BT$. Here, N represents the number of gas molecules in a system. Solving for N, we find

$$N = \frac{PV}{k_B T}$$

You complain: "But they didn't give me any information to plug in! It's hopeless!" Certainly not. The important thing to remember is that we have very little need for precision here. This is a rough estimate! *Just plug in a power of 10 for each variable.* Watch.

1. *Pressure*—Assume atmospheric pressure. Why not? Whether you're in Denver or Death Valley, the pressure in the room will be close to 10^5 N/m². (This value is on the constant sheet that is provided with the exam.)
2. *Volume*—They tell us, a cube 5 meters on a side. While this gives us 125 m³ as the volume of the room, we are going to round this off to 100 m³. This is close enough!
3. *Boltzmann's constant*—This is on the constant sheet. We only care about the order of magnitude; i.e., 10^{-23} J/K.
4. *Temperature*—The temperature of this classroom is unknown. Or is it? While we agree that the temperature of a classroom can vary from, say, 50–90°F, we only need to plug in the correct factor of 10 (in units of kelvins, of course). We propose that all classrooms in the known universe have a temperature above tens of kelvins, and below thousands of kelvins.[1] Room temperature is about 295 K. So let's just plug in 100 K for the temperature and be done with it.

Okay, we're ready for our quick calculation:

$$N = \frac{PV}{k_B T} = \frac{(10^5)(100)}{(10^{-23})(100)}$$

---

[1] Except for Mrs. Hawkins' fourth-grade class, which clearly existed within the jurisdictions of Hell, and thus was several orders of magnitude hotter.

The factors of 100 cancel, and we get that $N = 10^{28}$ molecules. (You remember that to divide powers of ten, just subtract the exponents.)

You still object: "But when I use my calculator and plug in the *precise* values, I get 3.271×10^{27} molecules." Look at the choices again. (C) is still the best answer; we got that without a calculator, and a lot quicker, too.

So what are the other ways of approaching the problem? One is to recognize that the room will contain not just one mole of gas, but some *thousands* of moles. You might remember from chemistry that one mole of an ideal gas takes up about twenty-two liters of space . . . and one mole of gas is ~10^{23} molecules. Thus, in this classroom, there must be 10^{23} molecules times many thousands. Of the choices, then, only 10^{28} molecules makes sense.

Or, you might remember or estimate that the distance between molecules of gases is usually in the neighborhood of 10^{-9} meters. If each side of the classroom is a few meters long, then the classroom is a cube with about 10^9 molecules on an edge. Thus, there are $(10^9)^3 = 10^{27}$ molecules in the whole room, and only choice (C) makes sense.

Proportional Reasoning

These also test your knowledge of how to use equations, except that you don't have to plug in numerical values to solve them.

> Planet X is twice as massive as Earth, but its radius is only half of Earth's radius. What is the acceleration due to gravity on Planet X in terms of g, the acceleration due to gravity on Earth?
>
> (A) $(^1/_4\, g)$
> (B) $(^1/_2\, g)$
> (C) g
> (D) $4\, g$
> (E) $8\, g$

First we need to know what equation to use. We know that the force that a planet exerts on a small mass m_1 near its surface is

$$F = \frac{GM_{planet}\, m_1}{R_{planet}^2}$$

Using Newton's second law ($F_{net} = ma$), we know that the acceleration of the small mass is simply

$$a = \frac{GM_{planet}}{R_{planet}^2}$$

One method of solution would be to plug in the actual mass and radius of the new planet. But no fair, you say, the mass of the Earth isn't given on the constants sheet. How do I find the mass of the planet?

You don't!

Use proportional reasoning skills instead, so:

"The mass of the planet is twice that of the Earth. Since mass is in the numerator of the equation for acceleration, doubling the mass of the planet must double the acceleration.

"Okay, but the radius of this planet is also different. Radius is in the denominator, so a smaller radius means a bigger acceleration. The radius of the new planet is half of the radius of the Earth. Therefore, the acceleration must be doubled. Almost there . . . because the radius is SQUARED in the denominator, the acceleration must be doubled AGAIN.

"So what is my final answer? The mass causes acceleration to double. The radius causes the acceleration to double, and then to double again. So the total acceleration is multiplied by a factor of 8. The acceleration on this planet is 8 g."

In the much more concise language of algebra, your reasoning might look like this:

$$a = \frac{2}{\left(\frac{1}{2}\right)^2} g = 8 g$$

What if the answer choices had been like this:

(A) 2.5 m/s^2
(B) 4.9 m/s^2
(C) 9.8 m/s^2
(D) 19.6 m/s^2
(E) 78.4 m/s^2

Is the problem any different? (Answer: no.)

Concept Questions: "WHY?"

Many multiple-choice questions involve no calculations and no formulas. These test your understanding of vocabulary and explanations for physical phenomena.

> At a smooth, sandy beach, it is observed that the water wave fronts are traveling almost perpendicular to the shoreline. However, a helicopter pilot notices that several hundred meters away from the shore, the wave fronts are traveling at about a 45° angle to the shore. What is a possible explanation for the change in the direction of the water wave fronts?
>
> (A) The water waves have a greater speed near the shore.
> (B) The water waves have a smaller speed near the shore.
> (C) The water waves have reflected off of the shore.
> (D) The acceleration due to gravity is greater near the shore.
> (E) The acceleration due to gravity is smaller near the shore.

The direct answer to this question is that if waves slow down or speed up, they bend. This is what is meant by *refraction* of waves. Because the waves are bending *toward* the shore, they must have slowed down. (Think about light waves bending when they go from a medium with low index of refraction to one with high index of refraction.) The answer is B.

But even if you are hesitant on questions about waves, you can still get close to the answer by eliminating obvious stupidicisms. Look at choices D and E. Now, we all know that the acceleration due to gravity is just about the same anywhere near the surface of the Earth. Why should *g* change just because waves are moving toward a shore? Eliminate D and E, and guess from the rest, if you have to.

Concept Questions: Diagrams

These ask you a simple question based (obviously) on a diagram.

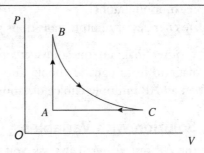

A gas undergoes the cyclic process represented by the pressure (P) vs. volume (V) diagram above. During which portion of this process is no work done on or by the gas?

(A) AB
(B) BC
(C) CA
(D) Some work is done on or by the gas in each of the three steps.
(E) No work is done on or by the gas in any of the three processes.

For these, you either know what to do with the diagram, or you don't. Here you, of course, remember that work done is given by the area under a curve on a PV diagram. You notice that there is no area under the vertical line AB; thus, the answer is (A).

> **TIP**
>
> ### Three Things You Can Do with a Graph
> You could see so, so many graphs on the AP exam . . . It's often difficult to remember which graph means what. But if you know your equations, you can usually figure out how to interpret any graph you are faced with. Why? Because there are pretty much ONLY three things you can do with a graph:
>
> 1. Take the slope.
> 2. Find the area under the graph.
> 3. Read off an axis.
>
> For example, one recent AP Physics B exam question showed a plot of energy vs. frequency, and asked for an experimental value of Planck's constant h. Chances are you have never done such an experiment, and that you've never seen this particular type of graph. But, you do remember your equations—the energy of a photon is related to frequency by $E = hf$. Solving for h, $h = \frac{E}{f}$. So h must be **slope** of the energy vs. frequency graph.
>
>   Similarly, imagine a graph of force vs. time on a question that asks for impulse. Since impulse is equal to force times time interval ($\Delta p = F\Delta t$), then impulse must be the **area** under the graph.
>
>   Finally, if you're totally clueless about what to do with a graph, just try taking a slope or an area, and see what happens! You might experience a revelation.

Other diagram questions might ask you to:

- use the right-hand rule to determine the direction of a magnetic force on a particle.
- identify the direction of an electric or magnetic field.

- analyze the properties of a circuit.
- use the sketch of a ray diagram to locate images in lenses and mirrors.
- recognize the correct free-body diagram of an object.
- interpret motion graphs.
- calculate energies of photon transitions given an atomic energy level diagram.

Many other diagram questions are possible. Try making one yourself—pick your favorite diagram from your textbook, and ask a question about it. Chances are, you have just written an AP multiple-choice question.

Direct Solution with Variables

Because the AP test writers can't ask you to do any kind of difficult number crunching on the multiple-choice section, often they will ask you to do your problem-solving using variables only.

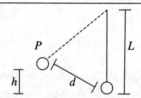

A pendulum of length L is drawn back to position P, as shown in the above diagram, and released from rest. The linear distance from P to the lowest point in the pendulum's swing is d; the vertical distance from P to the lowest point in the swing is h. What is the maximum speed of this pendulum in terms of the above variables and fundamental constants?

(A) $\sqrt{2gL}$

(D) $\sqrt{\dfrac{2gd}{L}}$

(B) $\sqrt{2gd}$

(E) $\sqrt{\dfrac{2gh}{L}}$

(C) $\sqrt{2gh}$

"Ugh . . . too many letters!" you say. We disagree. Solving this problem is no different than solving the same problem with numbers given. In fact, if the variables bother you, try solving with made-up numbers first:

> Let's say the height h is 5 meters, and the mass of the bob is 2 kg . . . well, we use conservation of energy. Energy at the top of the swing is all potential, all mgh. So that's $2 \times 10 \times 5 = 100$ J of potential energy.
>
> At the bottom, all this energy is kinetic. So 100 J $= \frac{1}{2}mv^2$. Solving, $v = 10$ m/s.
>
> Now how did we get that? We set $mgh = \frac{1}{2}mv^2$, and solved for v. The masses cancelled, so $v =$ square root of $2gh$. Lo and behold, that's an answer choice!

When Should You Skip a Question?

Never. There is no penalty for guessing, so guess away!

Some Final Advice on Multiple-Choice Questions

- Know your pace. Take the practice exams under test conditions (90 minutes for 70 problems, or some fraction thereof). Are you getting to all the questions? If not, you are going to need to decide your strengths and weaknesses. Know before the exam which types of problems you want to attempt first. Then, when you take your exam, FOLLOW YOUR PLAN!
- The multiple-choice questions do not necessarily start easy and get harder, as do SAT questions. So if you suspect from your practice that you may be pressed for time, know that problems on your strong topics may be scattered throughout the exam. Problem 70 might be easier for you than problem 5, so look at the whole test.
- Speaking of time, the AP test authors know the time limit of the exam—you must average a minute and a half per question in order to answer everything. So they are not going to write a question that really takes three or four minutes to solve! You must always look for the approach to a problem that will let you solve quickly. If your approach won't get you to a solution in less than two minutes, then either look for another approach or move on.
- One other alternative if you don't see a reasonable direct approach to a problem: look at the answer choices. Some might not make any sense; for example, you can eliminate any choice for a speed that is faster than light, or a couple of answer choices to concept questions might contain obvious errors. Guess from the remaining choices, and move on.
- Correct your practice exam. For any mistakes, write out an explanation of the correct answer and why you got it wrong. Pledge to yourself that you will never make the same mistake twice.

How to Approach the Free-Response Section

The best thing about the free-response section of the AP exam is this: you've been preparing for it *all year long!* "Really?" you ask . . . "I don't remember spending much time preparing for it."

But think about the homework problems you've been doing throughout the year. Every week, you probably answer a set of questions, each of which might take a few steps to solve, and we bet that your teacher always reminds you to show your work. This sounds like the AP free-response section to us!

The key to doing well on the free-response section is to realize that, first and foremost, these problems test your *understanding* of physics. The purpose is not to see how good your algebra skills are, how many fancy-sounding technical terms you know, or how many obscure theories you can regurgitate. So all we're going to do in this chapter is give you a few suggestions about how, when you work through a free-response question, you can communicate to the AP graders that you understand the concepts being tested. If you can effectively communicate your understanding of physics, you will get a good score.

What Do the Graders Look For?

Before grading a single student's exam, the high school and college physics teachers who are responsible for scoring the AP free-response section make a "rubric" for each question. A rubric is a grading guide; it specifies exactly what needs to be included for an answer to receive full credit, and it explains how partial credit should be awarded.

For example, consider part of a free-response question:

A student pulls a 1.0-kg block across a table to the right, applying a force of 8.0 N. The coefficient of kinetic friction between the block and the table is 0.20. Assume the block is at rest when it begins its motion.

(a) Determine the force of friction experienced by the block.
(b) Calculate the speed of the block after 1.5 s.

Let's look just at part (b). What do you think the AP graders are looking for in a correct answer? Well, we know that the AP free-response section tests your understanding of physics. So the graders probably want to see that you know how to evaluate the forces acting on an object and how to relate those forces to the object's motion.

In fact, if part (b) were worth 4 points, the graders might award 1 point for each of these elements of your answer:

1. Applying Newton's second law, $F_{net} = ma$, to find the block's acceleration.
2. Recognizing that the net force is not 8.0 N, but rather is the force of the student minus the force of friction (which was found in [a]), 8.0 N − 2.0 N = 6.0 N.
3. Using a correct kinematics equation with correct substitutions to find the final velocity of the block; i.e., $v_f = v_o + at$, where $v_o = 0$ and $a = 6.0$ N/1.0 kg $= 6.0$ m/s².
4. Obtaining a correct answer with correct units, 9.0 m/s.

Now, we're not suggesting that you try to guess how the AP graders will award points for every problem. Rather, we want you to see that the AP graders care much more about your understanding of physics than your ability to punch numbers into your calculator. Therefore, you should care much more about demonstrating your understanding of physics than about getting the right final answer.

Partial Credit

Returning to part (b) from the example problem, it's obvious that you can get lots of partial credit even if you make a mistake or two. For example:

- If you forgot to include friction, and just set the student's force equal to *ma* and solved, you could still get two out of four points.
- If you solved part (a) wrong but still got a reasonable answer, say 4.5 N for the force of friction, and plugged that in correctly here, you would still get either 3 or 4 points in part (b)! Usually the rubrics are designed not to penalize you twice for a wrong answer. So if you get part of a problem wrong, but your answer is consistent with your previous work, you'll usually get full or close to full credit.
- That said, if you had come up with a 1000 N force of friction, which is clearly unreasonable, you probably will not get credit for a wrong but consistent answer, unless you clearly indicate the ridiculousness of the situation. You'll still get probably two points, though, for the correct application of principles!
- If you got the right answer using a shortcut—say, doing the calculation of the net force in your head—you would not earn full credit but you would at least get the correct answer point. However, if you did the calculation *wrong* in your head, then you would *not* get any credit—AP graders can read what's written on the test, but they're not allowed to read your mind. Moral of the story: Communicate with the readers so you are sure to get all the partial credit you deserve.
- Notice how generous the partial credit is. You can easily get two or three points without getting the right answer and 50–75% is in the 4–5 range when the AP test is scored!

You should also be aware of some things that will NOT get you partial credit:

- You will not get partial credit if you write multiple answers to a single question. If AP graders see that you've written two answers, they will grade the one that's wrong. In other words, you will lose points if you write more than one answer to a question, even if one of the answers you write is correct.
- You will not get partial credit by including unnecessary information. There's no way to get extra credit on a question, and if you write something that's wrong, you could lose points. Answer the question fully, then stop.

The Tools You Can Use

You can use a calculator on the free-response section of the AP exam. Most calculators are acceptable—scientific calculators, programmable calculators, graphing calculators. However, you cannot use a calculator with a QWERTY keyboard, and you'd probably be restricted from using any calculators that make noise or that print their answers onto paper.[2] You also cannot share a calculator with anyone during the exam.

The real question, though, is whether a calculator will really help you. The short answer is "Yes": You will be asked questions on the exam that require you to do messy calculations (for example, you might need to divide a number by π, or multiply something by Boltzmann's constant[3]). The longer answer, though, is "Yes, but it won't help very much." To see what we mean, look back at the hypothetical grading rubric for part (b) of the example problem we discussed earlier. Two of the four possible points are awarded for using the right equations, one point is awarded for finding the magnitude of a force using basic arithmetic, and the last point is awarded for solving a relatively simple equation. So you would get half-credit if you did no math at all, and you would get full credit just by doing some very elementary math. You probably wouldn't need to touch your calculator!

So definitely bring a calculator to the exam but don't expect that you'll be punching away at it constantly.

The other tool you can use on the free-response section is the equations sheet. You will be given a copy of this sheet in your exam booklet. It's a handy reference because it lists all the equations that you're expected to know for the exam.

However, the equations sheet can also be dangerous. Too often, students interpret the equations sheet as an invitation to stop thinking: "Hey, they tell me everything I need to know, so I can just plug-and-chug through the rest of the exam!" Nothing could be further from the truth.

First of all, you've already *memorized* the equations on the sheet. It might be reassuring to look up an equation during the AP exam, just to make sure that you've remembered it correctly. And maybe you've forgotten a particular equation, but seeing it on the sheet will jog your memory. This is exactly what the equations sheet is for, and in this sense, it's pretty nice to have around. But beware of the following:

- Don't look up an equation unless you know *exactly* what you're looking for. It might sound obvious, but if you don't know what you're looking for, you won't find it.

---

[2]Does anyone actually use printing calculators anymore?

[3]You will, of course, be given the table of information during the free-response portion of the AP exam. This is the same table of information that you are given during the multiple-choice section.

- Don't go fishing. If part of a free-response question asks you to find an object's momentum, and you're not sure how to do that, don't just rush to the equations sheet and search for every equation with a "*P*" in it.

Other Advice About the Free-Response Section

- Always show your work. If you use the correct equation to solve a problem but you plug in the wrong numbers, you will probably get partial credit, but if you just write down an incorrect answer, you will definitely get no partial credit.
- If you don't know precisely how to solve a problem, simply explain your thinking process to the grader. If a problem asks you to find the centripetal acceleration of a satellite orbiting a planet, for example, and you don't know what equations to use, you might write something like this: "The centripetal force points toward the center of the satellite's orbit, and this force is due to gravity. If I knew the centripetal force, I could then calculate the centripetal acceleration using Newton's second law." This answer might earn you several points, even though you didn't do a single calculation.
- However, don't write a book. Keep your answers succinct.
- Let's say that part (b) of a question requires you to use a value calculated in part (a). You didn't know how to solve part (a), but you know how to solve part (b). What should you do? We can suggest two options. First, make up a reasonable answer for part (a), and then use that answer for part (b). Or, set some variable equal to the answer from part (a) (write a note saying something like, "Let *v* be the velocity found in part [a]"). Then, solve part (b) in terms of that variable. Both of these methods should allow you to get partial or even full credit on part (b).
- If you make a mistake, cross it out. If your work is messy, circle your answer so that it's easy to find. Basically, make sure the AP graders know what you want them to grade and what you want them to ignore.
- If you're stuck on a free-response question, try another one. Question #6 might be easier for you than question #1. Get the easy points first, and then only try to get the harder points if you have time left over.
- Always remember to use units where appropriate.
- It may be helpful to include a drawing or a graph in your answer to a question, but make sure to label your drawings or graphs so that they're easy to understand.
- No free-response question should take you more than 15–17 minutes to solve. They're not designed to be outrageously difficult, so if your answer to a free-response problem is outrageously complicated, you should look for a new way to solve the problem, or just skip it and move on.

Lab Questions

It is all well and good to be able to solve problems and calculate quantities using the principles and equations you've learned. However, the true test of any physics theory is whether or not it WORKS.

The AP development committee is sending a message to students that laboratory work is an important aspect of physics. To truly understand physics, you must be able to design and analyze experiments. Thus, *each free-response section will contain at least one question that involves experiment design and analysis.*

Here's an example:

In the laboratory, you are given a metal block, about the size of a brick. You are also given a 2.0-m-long wooden plank with a pulley attached to one end. Your goal is to determine experimentally the coefficient of kinetic friction, μ_k, between the metal block and the wooden plank.

(a) From the list below, select the additional equipment you will need to do your experiment by checking the line to the left of each item. Indicate if you intend to use more than one of an item.

   _____ 200-g mass   _____ 10-g mass   _____ spring scale
   _____ motion detector   _____ balance   _____ meterstick
   _____ a toy bulldozer that moves at constant speed
   _____ string

(b) Draw a labeled diagram showing how the plank, the metal block, and the additional equipment you selected will be used to measure μ_k.

(c) Briefly outline the procedure you will use, being explicit about what measurements you need to make and how these measurements will be used to determine μ_k.

To answer a lab question, just follow these steps:

1. **Follow the directions.**

    Sounds simple, doesn't it? When the test says, "Draw a diagram," it means they want you to draw a diagram. And when it says, "Label your diagram," it means they want you to label your diagram. You will likely earn points just for these simple steps.

Exam tip from an AP Physics veteran:

On the AP test, I forgot to label point *B* on a diagram, even though I obviously knew where point *B* was. This little mistake cost me several points!

                                                  —*Zack, college senior and engineer*

---

2. **Use as few words as possible.**

    Answer the question, then stop. You can lose credit for an incorrect statement, even if the other 15 statements in your answer are correct. The best idea is to keep it simple.

3. **There is no single correct answer.**

    Most of the lab questions are open-ended. There might be four or more different correct approaches. So don't try to "give them the answer they're looking for." Just do something that seems to make sense—you might well be right!

4. **Don't assume you have to use all the stuff they give you.**

    It might sound fun to use a force probe while determining the index of refraction of a glass block, but, really! A force probe!?!

5. **Don't over-think the question.**

   They're normally not too complicated. Remember, you're supposed to take only 10–15 minutes to write your answer. You're not exactly designing a subatomic particle accelerator.

6. **Don't state the obvious.**

   You may assume that basic lab protocols will be followed. So there's no need to tell the reader that you recorded your data carefully, nor do you need to remind the reader to wear safety goggles.

Now Put It All Together

Here are two possible answers to the sample question. Look how explicit we were about what quantities are measured, how each quantity is measured, and how μ_k is determined. We aren't *artistes*, so our diagram doesn't look so good. But for the AP exam, we believe in substance over style. All the necessary components are there, and that's all that matters.

Answer #1

In the laboratory, you are given a metal block, about the size of a brick. You are also given a 2.0-m-long wooden plank with a pulley attached to one end. Your goal is to determine experimentally the coefficient of kinetic friction, μ_k, between the metal block and the wooden plank.

(a) From the list below, select the additional equipment you will need to do your experiment by checking the line to the left of each item. Indicate if you intend to use more than one of an item.

    _____ 200-g mass   _____ 10-g mass   ✔ spring scale

    _____ motion detector   ✔ balance   meterstick

    ✔ a toy bulldozer that moves at constant speed

    ✔ string

(b) Draw a labeled diagram showing how the plank, the metal block, and the additional equipment you selected will be used to measure μ_k.

(c) Briefly outline the procedure you will use, being explicit about what measurements you need to make and how these measurements will be used to determine μ_k.

Use the balance to determine the mass, m, of the metal block. The weight of the block is mg. Attach the spring scale to the bulldozer; attach the other end of the spring scale to the metal block with string. Allow the bulldozer to pull the block at constant speed.

The block is in equilibrium. So, the reading of the spring scale while the block is moving is the friction force on the block; the normal force on the block is equal to its weight. The coefficient of kinetic friction is equal to the spring scale reading divided by the block's weight.

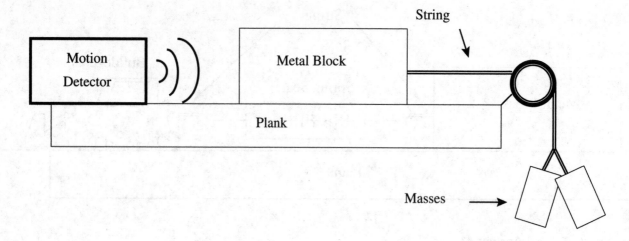

Answer #2

In the laboratory, you are given a metal block, about the size of a brick. You are also given a 2.0-m-long wooden plank with a pulley attached to one end. Your goal is to determine experimentally the coefficient of kinetic friction, μ_k, between the metal block and the wooden plank.

(a) From the list below, select the additional equipment you will need to do your experiment by checking the line to the left of each item. Indicate if you intend to use more than one of an item.

   ✔ 200-g mass   ✔ 10-g mass   _____ spring scale
   (several)   (several)

   ✔ motion detector   ✔ balance   _____ meterstick

   _____ a toy bulldozer that moves at constant speed

   ✔ string

(b) Draw a labeled diagram showing how the plank, the metal block, and the additional equipment you selected will be used to measure μ_k.

(c) Briefly outline the procedure you will use, being explicit about what measurements you need to make and how these measurements will be used to determine μ_k.

Determine the mass, m, of the block with the balance. The weight of the block is mg. Attach a string to the block and pass the string over the pulley. Hang masses from the other end of the string, changing the amount of mass until the block can move across the plank at constant speed. Use the motion detector to verify that the speed of the block is as close to constant as possible.

The block is in equilibrium. So, the weight of the hanging masses is equal to the friction force on the block; the normal force on the block is equal to its weight. The coefficient of kinetic friction is thus equal to the weight of the hanging masses divided by the block's weight.

CHAPTER 8

Extra Drill on Difficult but Frequently Tested Topics

IN THIS CHAPTER
Summary: Drills in five types of problems that you should spend extra time reviewing, with step-by-step solutions.

Key Ideas
- ✪ Tension problems
- ✪ Electric and magnetic fields problems
- ✪ Inclined plane problems
- ✪ Motion graph problems
- ✪ Simple circuits problems

Practice problems and tests cannot possibly cover every situation that you may be asked to understand in physics. However, some categories of topics come up again and again, so much so that they might be worth some extra review. And that's exactly what this chapter is for—to give you a focused, intensive review of a few of the most essential physics topics.

We call them "drills" for a reason. They are designed to be skill-building exercises, and as such, they stress repetition and technique. Working through these exercises might remind you of playing scales if you're a musician or of running laps around the field if you're an athlete. Not much fun, maybe a little tedious, but very helpful in the long run.

The questions in each drill are all solved essentially the same way. *Don't* just do one problem after the other . . . rather, do a couple, check to see that your answers are right,[1] and then, half an hour or a few days later, do a few more, just to remind yourself of the techniques involved.

---

[1]For numerical answers, it's okay if you're off by a significant figure or so.

Tension

How to Do It

Use the following steps to solve these kinds of problems: (1) Draw a free-body diagram for each block; (2) resolve vectors into their components; (3) write Newton's second law for each block, being careful to stick to your choice of positive direction; and (4) solve the simultaneous equations for whatever the problem asks for.

The Drill

In the diagrams below, assume all pulleys and ropes are massless, and use the following variable definitions.

$$F = 10 \text{ N}$$
$$M = 1.0 \text{ kg}$$
$$\mu = 0.2$$

Find the tension in each rope and the acceleration of the set of masses.
(For a greater challenge, solve in terms of F, M, and μ instead of plugging in values.)

1. Frictionless

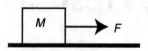

2. Frictionless

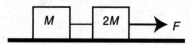

3. Frictionless

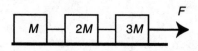

4. Coefficient of Friction μ

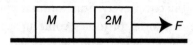

5.

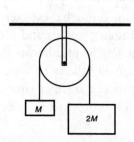

6.

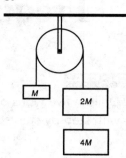

7. Frictionless

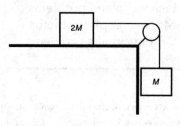

8. Frictionless

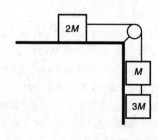

9. Frictionless

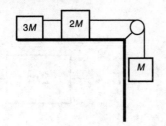

10. Coefficient of Friction μ

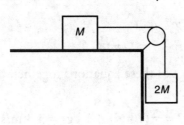

11. Coefficient of Friction μ

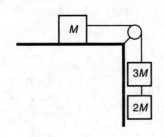

12. Frictionless

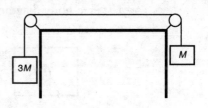

13. Frictionless

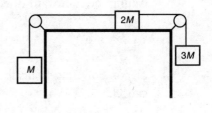

14. Coefficient of Friction μ

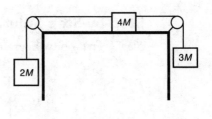

› The Answers
(Step-by-Step Solutions to #2 and #5 Are on the Next Page.)

1. $a = 10$ m/s^2

2. $a = 3.3$ m/s^2
$T = 3.3$ N

3. $a = 1.7$ m/s^2
$T_1 = 1.7$ N
$T_2 = 5.1$ N

4. $a = 1.3$ m/s^2
$T = 3.3$ N

5. $a = 3.3$ m/s^2
$T = 13$ N

6. $a = 7.1$ m/s^2
$T_1 = 17$ N
$T_2 = 11$ N

7. $a = 3.3$ m/s^2
$T = 6.6$ N

8. $a = 6.7$ m/s^2
$T_1 = 13$ N
$T_2 = 10$ N

9. $a = 1.7$ m/s^2
$T_1 = 5.1$ N
$T_2 = 8.3$ N

10. $a = 6.0$ m/s^2
$T = 8.0$ N

11. $a = 8.0$ m/s^2
$T_1 = 10$ N
$T_2 = 4.0$ N

12. $a = 5.0$ m/s^2
$T = 15$ N

13. $a = 3.3$ m/s^2
$T_1 = 13$ N
$T_2 = 20$ N

14. $a = 0.22$ m/s^2
$T_1 = 20$ N
$T_2 = 29$ N

Step-by-Step Solution to #2:

Step 1: Free-body diagrams:

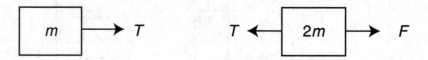

No components are necessary, so on to *Step 3:* write Newton's second law for each block, calling the rightward direction positive:

$$T - 0 = ma$$
$$F - T = (2m)a$$

Step 4: Solve algebraically. It's easiest to add these equations together, because the tensions cancel:

$$F = (3m)a, \text{ so } a = F/3m = (10 \text{ N})/3(1 \text{ kg}) = 3.3 \text{ m/s}^2$$

To get the tension, just plug back into $T - 0 = ma$ to find $T = F/3 = 3.3$ N.

Step-by-Step Solution to #5:

Step 1: Free-body diagrams:

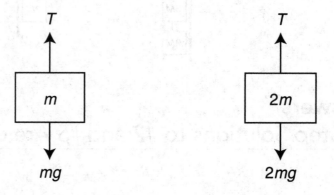

No components are necessary, so on to *Step 3:* write Newton's second law for each block, calling clockwise rotation of the pulley positive:

$$(2m)g - T = (2m)a$$
$$T - mg = ma$$

Step 4: Solve algebraically. It's easiest to add these equations together, because the tensions cancel:

$$mg = (3m)a, \text{ so } a = g/3 = 3.3 \text{ m/s}^2$$

To get the tension, just plug back into $T - mg = ma$: $T = m(a + g) = (4/3)mg = 13$ N.

Electric and Magnetic Fields

How to Do It

The force of an electric field is $F = qE$, and the direction of the force is in the direction of the field for a positive charge. The force of a magnetic field is $F = qvB\sin\theta$, and the direction of the force is given by the right-hand rule.

The Drill

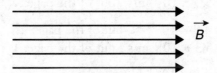

The magnetic field above has magnitude 3.0 T. For each of the following particles placed in the field, find (a) the force exerted by the magnetic field on the particle, and (b) the acceleration of the particle. Be sure to give magnitude *and direction* in each case.

1. an e^- at rest
2. an e^- moving ↑ at 2 m/s
3. an e^- moving ← at 2 m/s
4. a proton moving out of the page at 2 m/s
5. an e^- moving up and to the right, at an angle of 30° to the horizontal, at 2 m/s
6. an e^- moving up and to the left, at an angle of 30° to the horizontal, at 2 m/s
7. a positron moving up and to the right, at an angle of 30° to the horizontal, at 2 m/s
8. an e^- moving → at 2 m/s
9. a proton moving into the page at 2 m/s

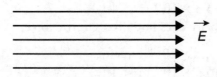

The electric field above has magnitude 3.0 N/C. For each of the following particles placed in the field, find (a) the force exerted by the electric field on the particle, and (b) the acceleration of the particle. Be sure to give magnitude *and direction* in each case.

10. an e^- at rest
11. a proton at rest
12. a positron at rest
13. an e^- moving ↑ at 2 m/s
14. an e^- moving → at 2 m/s
15. a proton moving out of the page at 2 m/s
16. an e^- moving ← at 2 m/s
17. a positron moving up and to the right, at an angle of 30° to the horizontal, at 2 m/s

› The Answers
(Step-by-Step Solutions to #2 and #10 Are on the Next Page.)

1. No force or acceleration, $v = 0$.

2. $F = 9.6 \times 10^{-19}$ N, out of the page.
   $a = 1.1 \times 10^{12}$ m/s^2, out of the page.

3. No force or acceleration, $\sin \theta = 0$.

4. $F = 9.6 \times 10^{-19}$ N, toward the top of the page.
   $a = 5.6 \times 10^8$ m/s^2, toward the top of the page.

5. $F = 4.8 \times 10^{-19}$ N, out of the page.
   $a = 5.3 \times 10^{11}$ m/s^2, out of the page.

6. $F = 4.8 \times 10^{-19}$ N, out of the page.
   $a = 5.3 \times 10^{11}$ m/s^2, out of the page.

7. $F = 4.8 \times 10^{-19}$ N, into the page.
   $a = 5.3 \times 10^{11}$ m/s^2, into the page.

8. No force or acceleration, $\sin \theta = 0$.

9. $F = 9.6 \times 10^{-19}$ N, toward the bottom of the page.
   $a = 5.6 \times 10^{8}$ m/s^2, toward the bottom of the page.

10. $F = 4.8 \times 10^{-19}$ N, left.
    $a = 5.3 \times 10^{11}$ m/s^2, left.

11. $F = 4.8 \times 10^{-19}$ N, right.
    $a = 2.8 \times 10^{8}$ m/s^2, right.

12. $F = 4.8 \times 10^{-19}$ N, right.
    $a = 5.3 \times 10^{11}$ m/s^2, right.

13. $F = 4.8 \times 10^{-19}$ N, left.
    $a = 5.3 \times 10^{11}$ m/s^2, left.
    Velocity does not affect electric force.

14. $F = 4.8 \times 10^{-19}$ N, left.
    $a = 5.3 \times 10^{11}$ m/s^2, left.

15. $F = 4.8 \times 10^{-19}$ N, right.
    $a = 2.8 \times 10^{8}$ m/s^2, right.

16. $F = 4.8 \times 10^{-19}$ N, left.
    $a = 5.3 \times 10^{11}$ m/s^2, left.

17. $F = 4.8 \times 10^{-19}$ N, right.
    $a = 5.3 \times 10^{11}$ m/s^2, right.

Step-by Step Solution to #2:

(a) The magnetic force on a charge is given by $F = qvB \sin \theta$. Since the velocity is perpendicular to the magnetic field, $\theta = 90°$, and $\sin \theta = 1$. The charge q is the amount of charge on an electron, 1.6×10^{-19} C. v is the electron's speed, 2 m/s. B is the magnetic field, 3 T.

$$F = (1.6 \times 10^{-19} \text{ C})(2 \text{ m/s})(3 \text{ T})(1) = 9.6 \times 10^{-19} \text{ N}$$

The direction is given by the right-hand rule. Point your fingers in the direction of the electron's velocity, toward the top of the page; curl your fingers in the direction of the magnetic field, to the right; your thumb points into the page. Since the electron has a negative charge, the force points opposite your thumb, or out of the page.

(b) Even though we're dealing with a magnetic force, we can still use Newton's second law. Since the magnetic force is the only force acting, just set this force equal to ma and solve. The direction of the acceleration must be in the same direction as the net force.

$$9.6 \times 10^{-19} \text{ N} = (9.1 \times 10^{-31} \text{ kg})a$$
$$a = 1.1 \times 10^{12} \text{ m/s}^2, \text{ out of the page}$$

Step-by-Step Solution to #10:

(a) The electric force on a charge is given by $F = qE$. The charge q is the amount of charge on an electron, 1.6×10^{-19} C. E is the electric field, 3 N/C.

$$F = (1.6 \times 10^{-19} \text{ C})(3 \text{ N/C}) = 4.8 \times 10^{-19} \text{ N}$$

Because the electron has a negative charge, the force is opposite the electric field, or left.

(b) Even though we're dealing with an electric force, we can still use Newton's second law. Since the electric force is the only force acting, just set this force equal to ma and solve. The direction of the acceleration must be in the same direction as the net force.

$$4.8 \times 10^{-19} \text{ N} = (9.1 \times 10^{-31} \text{ kg})a$$
$$a = 2.8 \times 10^{8} \text{ m/s}^2, \text{ left}$$

Inclined Planes

How to Do It

Use the following steps to solve these kinds of problems: (1) Draw a free-body diagram for the object (the normal force is perpendicular to the plane; the friction force acts along the plane, opposite the velocity); (2) break vectors into components, where the parallel component of weight is $mg(\sin \theta)$; (3) write Newton's second law for parallel and perpendicular components; and (4) solve the equations for whatever the problem asks for.

Don't forget, the normal force is NOT equal to mg when a block is on an incline!

The Drill

Directions: For each of the following situations, determine:

(a) the acceleration of the block down the plane
(b) the time for the block to slide to the bottom of the plane

In each case, assume a frictionless plane unless otherwise stated; assume the block is released from rest unless otherwise stated.

1.

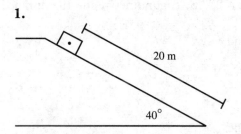

2.

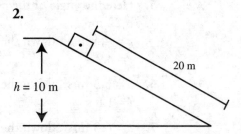

3.

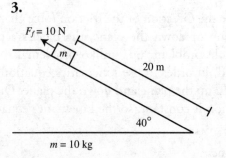

4.

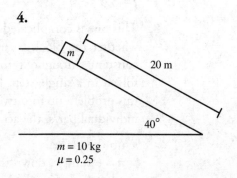

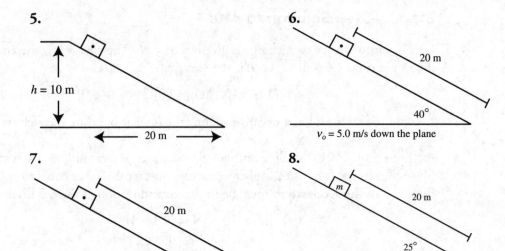

5.

$h = 10$ m

20 m

6.

20 m

40°

$v_0 = 5.0$ m/s down the plane

7.

20 m

40°

$v_0 = 5.0$ m/s up the plane

8.

m

20 m

25°

$m = 30$ kg
$\mu = 0.30$
$v_0 = 3.0$ m/s up the plane

Careful—this one's tricky.

> The Answers
(A Step-by-Step Solution to #1 Is on the Next Page.)

1. $a = 6.3$ m/s², down the plane.
    $t = 2.5$ s

2. $a = 4.9$ m/s², down the plane.
    $t = 2.9$ s

3. $a = 5.2$ m/s², down the plane.
    $t = 2.8$ s

4. $a = 4.4$ m/s², down the plane.
    $t = 3.0$ s

5. Here the angle of the plane is 27° by trigonometry, and the distance along the plane
    is 22 m.
    $a = 4.4$ m/s², down the plane.
    $t = 3.2$ s

6. $a = 6.3$ m/s², down the plane.
    $t = 1.8$ s

7. $a = 6.3$ m/s², down the plane.
    $t = 3.5$ s

8. This one is complicated. Since the direction of the friction force changes depending on
    whether the block is sliding up or down the plane, the block's acceleration is NOT
    constant throughout the whole problem. So, unlike problem #7, this one can't be
    solved in a single step. Instead, in order to use kinematics equations, you must break
    this problem up into two parts: up the plane and down the plane. During each of these
    individual parts, the acceleration is constant, so the kinematics equations are valid.

    - up the plane:
      $a = 6.8$ m/s², down the plane.
      $t = 0.4$ s before the block turns around to come down the plane.

- down the plane:
  $a = 1.5$ m/s^2, down the plane.
  $t = 5.2$ s to reach bottom.

So, a total of $t = 5.6$ s for the block to go up and back down.

Step-by-Step Solution to #1:

Step 1: Free-body diagram:

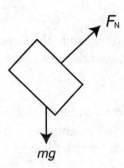

Step 2: Break vectors into components. Because we have an incline, we use inclined axes, one parallel and one perpendicular to the incline:

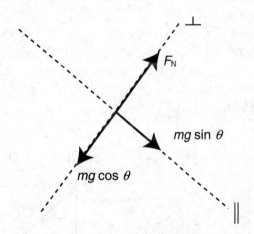

Step 3: Write Newton's second law for each axis. The acceleration is entirely directed parallel to the plane, so perpendicular acceleration can be written as zero:

$$mg \sin \theta - 0 = ma$$
$$F_N - mg \cos \theta = 0$$

Step 4: Solve algebraically for a. This can be done without reference to the second equation. (In problems with friction, use $F_f = \mu F_N$ to relate the two equations.)

$$a = g \sin \theta = 6.3 \text{ m/s}^2$$

To find the time, plug into a kinematics chart:

$$v_o = 0$$
$$v_f = \text{unknown}$$
$$\Delta x = 20 \text{ m}$$
$$a = 6.3 \text{ m/s}^2$$
$$t = ???$$

Solve for t using the second star equation for kinematics (**): $\Delta x = v_o t + \frac{1}{2}at^2$, where v_o is zero:

$$t = \sqrt{\frac{2\Delta x}{a}} = \sqrt{\frac{2(20 \text{ m})}{6.3 \text{ m/s}^2}} = 2.5 \text{ s}$$

Motion Graphs

How to Do It

For a position–time graph, the slope is the velocity. For a velocity–time graph, the slope is the acceleration, and the area under the graph is the displacement.

The Drill

Use the graph to determine something about the object's speed. Then play "Physics *Taboo*": suggest what object might reasonably perform this motion and explain in words how the object moves. Use everyday language. In your explanation, you may *not* use any words from the list below:

velocity
acceleration
positive
negative
increase
decrease
it
object
constant

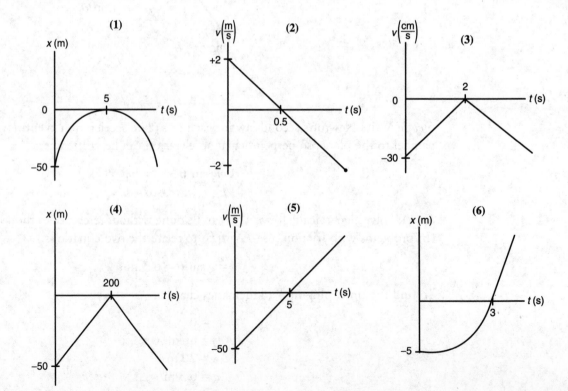

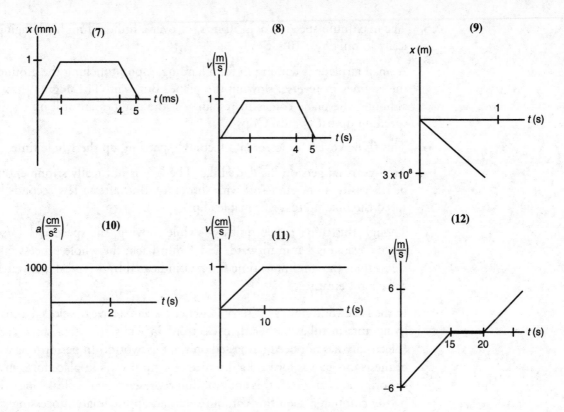

› The Answers

Note that our descriptions of the moving objects reflect our own imaginations. You might have come up with some very different descriptions, and that's fine ... provided that your answers are conceptually the same as ours.

1. The average speed over the first 5 s is 10 m/s, or about 22 mph. So:

    Someone rolls a bowling ball along a smooth road. When the graph starts, the bowling ball is moving along pretty fast, but the ball encounters a long hill. So, the ball slows down, coming to rest after 5 s. Then, the ball comes back down the hill, speeding up the whole way.

2. This motion only lasts 1 s, and the maximum speed involved is about 5 mph. So:

    A biker has been cruising up a hill. When the graph starts, the biker is barely moving at jogging speed. Within half a second, and after traveling only a meter up the hill, the bike turns around, speeding up as it goes back down the hill.

3. The maximum speed of this thing is 30 cm/s, or about a foot per second. So:

    A toy racecar is moving slowly along its track. The track goes up a short hill that's about a foot long. After 2 s, the car has just barely reached the top of the hill, and is perched there momentarily; then, the car crests the hill and speeds up as it goes down the other side.

4. The steady speed over 200 s (a bit over 3 minutes) is 0.25 m/s, or 25 cm/s, or about a foot per second.

    A cockroach crawls steadily along the school's running track, searching for food. The cockroach starts near the 50 yard line of the football field; around three minutes later, the cockroach reaches the goal line and, having found nothing of interest, turns around and crawls at the same speed back toward his starting point.

5. The maximum speed here is 50 m/s, or over a hundred mph, changing speed dramatically in only 5 or 10 s. So:

   A small airplane is coming in for a landing. Upon touching the ground, the pilot puts the engines in reverse, slowing the plane. But wait! The engine throttle is stuck! So, although the plane comes to rest in 5 s, the engines are still on . . . the plane starts speeding up backwards! Oops . . .

6. This thing covers 5 meters in 3 seconds, speeding up the whole time.

   An 8-year-old gets on his dad's bike. The boy is not really strong enough to work the pedals easily, so he starts off with difficulty. But, after a few seconds he's managed to speed the bike up to a reasonable clip.

7. Though this thing moves quickly—while moving, the speed is 1 m/s—the total distance covered is 1 mm forward, and 1 mm back; the whole process takes 5 ms, which is less than the minimum time interval indicated by a typical stopwatch. So we'll have to be a bit creative:

   In the Discworld novels by Terry Pratchett, wizards have developed a computer in which living ants in tubes, rather than electrons in wires and transistors, carry information. (Electricity has not been harnessed on the Discworld.) In performing a calculation, one of these ants moves forward a distance of 1 mm; stays in place for 3 ms; and returns to the original position. If this ant's motion represents two typical "operations" performed by the computer, then this computer has an approximate processing speed of 400 Hz times the total number of ants inside.

8. Though this graph *looks* like #7, this one is a velocity–time graph, and so indicates completely different motion.

   A small child pretends he is a bulldozer. Making a "brm-brm-brm" noise with his lips, he speeds up from rest to a slow walk. He walks for three more seconds, then slows back down to rest. He moved forward the entire time, traveling a total distance (found from the area under the graph) of 4 m.

9. This stuff moves 300 million meters in 1 s at a constant speed. There's only one possibility here: electromagnetic waves in a vacuum.

   Light (or electromagnetic radiation of any frequency) is emitted from the surface of the moon. In 1 s, the light has covered about half the distance to Earth.

10. Be careful about axis labels: this is an *acceleration*–time graph. Something is accelerating at 1000 cm/s^2 for a few seconds. 1000 cm/s^2 = 10 m/s^2, about Earth's gravitational acceleration. Using kinematics, we calculate that if we drop something from rest near Earth, after 4 s the thing has dropped 80 m.

    One way to simulate the effects of zero gravity is to drop an experiment from the top of a high tower. Then, because everything that was dropped is speeding up at the same rate, the effect is just as if the experiment were done in the Space Shuttle—at least until everything hits the ground. In this case, an experiment is dropped from a 250-ft tower, hitting the ground with a speed close to 90 mph.

11. 1 cm/s is ridiculously slow. Let's use the world of slimy animals:

    A snail wakes up from his nap and decides to find some food. He speeds himself up from rest to his top speed in 10 s. During this time, he's covered 5 cm, or about the length of your pinkie finger. He continues to slide along at a steady 1 cm/s, which means that a minute later he's gone no farther than a couple of feet. Let's hope that food is close.

12. This one looks a bit like those up-and-down-a-hill graphs, but with an important difference—this time the thing stops not just for an instant, but for five whole seconds, before continuing back toward the starting point.

A bicyclist coasts to the top of a shallow hill, slowing down from cruising speed (~15 mph) to rest in 15 s. At the top, she pauses briefly to turn her bike around; then, she releases the brake and speeds up as she goes back down the hill.

Simple Circuits

How to Do It

Think "series" and "parallel." The current through series resistors is the same, and the voltage across series resistors adds to the total voltage. The current through parallel resistors adds to the total current, and the voltage across parallel resistors is the same.

The Drill

For each circuit drawn below, find the current through and voltage across each resistor.
　　Note: Assume each resistance and voltage value is precise to two significant figures.

1.

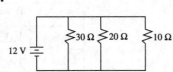

2.

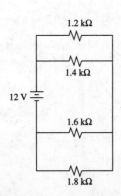

3.

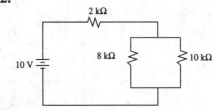

4.

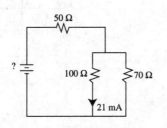

5.

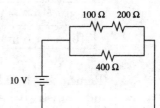

6.

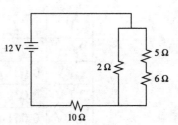

7.

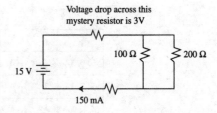

❯ The Answers
(A Step-by-Step Solution to #2 Is on the Next Page.)

1.

V	I	R
12 V	0.40 A	30 Ω
12 V	0.60 A	20 Ω
12 V	1.2 A	10 Ω
12 V	2.2 A	5.5 Ω

2.

V	I	R
3.2 V	1.6 mA	2 kΩ
6.8 V	0.9 mA	8 kΩ
6.8 V	0.7 mA	10 kΩ
10 V	1.6 mA	6.4 kΩ

(Remember, a kΩ is 1000 Ω, and a mA is 10^{-3} A.)

3.

V	I	R
2.5 V	0.051 A	50 Ω
2.1 V	0.021 A	100 Ω
2.1 V	0.030 A	70 Ω
4.6 V	0.051 A	91 Ω

4.

V	I	R
5.2 V	4.3 mA	1.2 kΩ
5.2 V	3.7 mA	1.4 kΩ
6.8 V	4.2 mA	1.6 kΩ
6.8 V	3.8 mA	1.8 kΩ
12 V	8.0 mA	1.5 kΩ

5.

V	I	R
3.4 V	0.034 A	100 Ω
6.8 V	0.034 A	200 Ω
10 V	0.025 A	400 Ω
10 V	0.059 A	170 Ω

6.

V	I	R
1.8 V	0.90 A	2.0 Ω
0.7 V	0.13 A	5.0 Ω
0.8 V	0.13 A	6.0 Ω
10.3 V	1.03 A	10.0 Ω
12.0 V	1.03 A	11.7 Ω

7.

V	I	R
3 V	0.15 A	20 Ω
10 V	0.10 A	100 Ω
10 V	0.05 A	200 Ω
2 V	0.15 A	13 Ω
15 V	0.15 A	100 Ω

Step-by-Step Solution to #2:

Start by simplifying the combinations of resistors. The 8 kΩ and 10 kΩ resistors are in parallel. Their equivalent resistance is given by

$$\frac{1}{R_{eq}} = \frac{1}{8\,k\Omega} + \frac{1}{10\,k\Omega}$$

which gives $R_{eq} = 4.4$ kΩ.

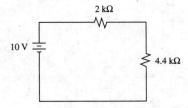

Next, simplify these series resistors to their equivalent resistance of 6.4 kΩ.

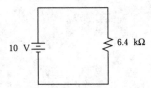

6.4 kΩ (i.e., 6400 Ω) is the total resistance of the entire circuit. Because we know the total voltage of the entire circuit to be 10 V, we can use Ohm's law to get the total current

$$I_{total} = \frac{V_{total}}{R_{total}} = \frac{10\,V}{6400\,\Omega} = 0.0016\,A$$

(more commonly written as 1.6 mA).

Now look at the previous diagram. The same current of 1.6 mA must go out of the battery, into the 2 kΩ resistor, and into the 4.4 kΩ resistor. The voltage across each resistor can thus be determined by $V = (1.6\,mA)R$ for each resistor, giving 3.2 V across the 2 kΩ resistor and 6.8 V across the 4.4 kΩ resistor.

The 2 kΩ resistor is on the chart. However, the 4.4 kΩ resistor is the equivalent of two parallel resistors. Because voltage is the same for resistors in parallel, there are 6.8 V across *each* of the two parallel resistors in the original diagram. Fill that in the chart, and use Ohm's law to find the current through each:

$$I_{8k} = 6.8\,V/8000\,\Omega = 0.9\,mA$$
$$I_{10k} = 6.8\,V/10,000\,\Omega = 0.7\,mA$$

STEP 4

Review the Knowledge You Need to Score High

CHAPTER 9

A Bit About Vectors

IN THIS CHAPTER

Summary: Understand the difference between scalars and vectors, how to draw vectors, how to break down vectors into components, and how to add vectors.

Key Ideas

✪ Scalars are quantities that have a magnitude but no direction—for example, temperature; in contrast, vectors have both magnitude *and* direction—for example, velocity.

✪ Vectors are drawn as arrows; the length of the arrow corresponds to the magnitude of the vector, and the direction of the arrow represents the direction of the vector.

✪ Any vector can be broken down into its *x*- and *y*-components; breaking a vector into its components will make many problems simpler.

Relevant Equations

Components of a vector V: $V_x = v\cos\theta$
$V_y = v\sin\theta$

Note: this assumes that θ is measured from the horizontal. These equations are not on the equation sheet, but should be memorized.

Scalars and vectors are easy. So we'll make this quick.

Scalars

Scalars are numbers that have a magnitude but no direction.

Magnitude: How big something is

For example, temperature is a scalar. On a cold winter day, you might say that it is "4 degrees" outside. The units you used were "degrees." But the temperature was not oriented in a particular way; it did not have a direction.

Another scalar quantity is speed. While traveling on a highway, your car's speedometer may read "70 miles per hour." It does not matter whether you are traveling north or south, if you are going forward or in reverse: your speed is 70 miles per hour.

Vector Basics

Vectors, by comparison, have both magnitude *and* direction.

> **Direction:** The orientation of a vector

An example of a vector is velocity. Velocity, unlike speed, always has a direction. So, let's say you are traveling on the highway again at a speed of 70 miles per hour. First, define what direction is positive—we'll call north the positive direction. So, if you are going north, your velocity is +70 miles per hour. The magnitude of your velocity is "70 miles per hour," and the direction is "north."

If you turn around and travel south, your velocity is −70 miles per hour. The magnitude (the speed) is still the same, but the sign is reversed because you are traveling in the negative direction. The direction of your velocity is "south."

 IMPORTANT: If the answer to a free-response question is a vector quantity, be sure to state *both* the magnitude and direction. However, don't use a negative sign if you can help it! Rather than "−70 miles per hour," state the true meaning of the negative sign: "70 miles per hour, south."

Graphic Representation of Vectors

Vectors are drawn as arrows. The length of the arrow corresponds to the magnitude of the vector—the longer the arrow, the greater the magnitude of the vector. The direction in which the arrow points represents the direction of the vector. Figure 9.1 shows a few examples:

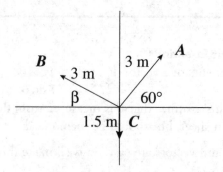

Figure 9.1   Examples of vectors.

Vector **A** has a magnitude of 3 meters. Its direction is "60 degrees above the positive *x*-axis." Vector **B** also has a magnitude of 3 meters. Its direction is "β degrees above the negative *x*-axis." Vector **C** has a magnitude of 1.5 meters. Its direction is "in the negative *y*-direction" or "90 degrees below the *x*-axis."

Vector Components

Any vector can be broken into its *x*- and *y*-components. Here's what we mean:

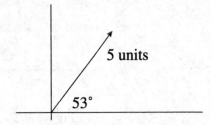

Figure 9.2   Vector components.

Place your finger at the tail of the vector in Figure 9.2 (that's the end of the vector that does not have a ◄ on it). Let's say that you want to get your finger to the head of the vector without moving diagonally. You would have to move your finger three units to the right and four units up. Therefore, the magnitude of left–right component (*x*-component) of the vector is "3 units" and the magnitude of up–down (*y*-component) of the vector is "4 units."

If your languages of choice are Greek and math, then you may prefer this explanation.

> Given a vector **V** with magnitude *v* directed at an angle θ above the horizontal
>
> $$V_x = v \cos \theta$$
> $$V_y = v \sin \theta$$

You may want to check to see that these formulas work by plugging in the values from our last example.

$$V_x = 5 \cos 53° = 3 \text{ units}$$
$$V_y = 5 \sin 53° = 4 \text{ units}$$

> **Exam tip from an AP Physics veteran:**
> Even though the vector formulas in the box are not on the equations sheet, they are very important to memorize. You will use them in countless problems. Chances are, you will use them so much that you'll have memorized them way before the AP exam.
>
> —*Jamie, high school senior*

Adding Vectors

Let's take two vectors, **Q** and **Z**, as shown in Figure 9.3a.

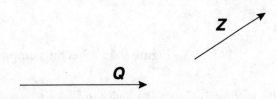

Figure 9.3a   Adding vectors.

Now, in Figure 9.3b, we place them on a coordinate plane. We will move them around so that they line up head-to-tail.

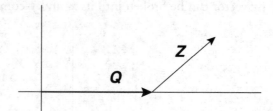

Figure 9.3b Adding vectors.

If you place your finger at the origin and follow the arrows, you will end up at the head of vector **Z**. The vector sum of **Q** and **Z** is the vector that starts at the origin and ends at the head of vector **Z**. This is shown in Figure 9.3c.

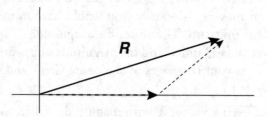

Figure 9.3c Adding vectors.

Physicists call the vector sum the "resultant vector." Usually, we prefer to call it "the resultant" or, as in our diagram, "**R**."

How to add vectors:
1. Line them up head-to-tail.
2. Draw a vector that connects the tail of the first arrow to the head of the last arrow.

Vector Components, Revisited

Breaking a vector into its components will make many problems simpler. Here's an example:

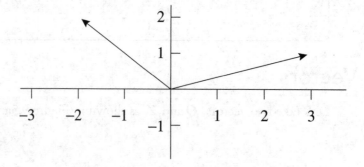

Figure 9.4a Vector components.

To add the vectors in Figure 9.4a, all you have to do is add their *x*- and *y*-components. The sum of the *x*-components is 3 + (−2) = 1 units. The sum of the *y*-components is

1 + 2 = 3 units. The resultant vector, therefore, has an *x*-component of +1 units and a *y*-component of +3 units. See Figure 9.4b.

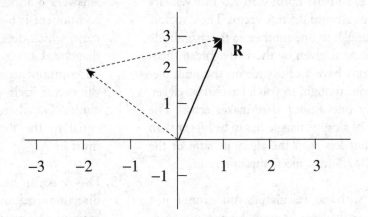

Figure 9.4b   Vector components.

Some Final Hints

1. Make sure your calculator is set to DEGREES, not radians.
2. Always use units. Always. We mean it. Always.

› Practice Problems

1. A canoe is paddled due north across a lake at 2.0 m/s relative to still water. The current in the lake flows toward the east; its speed is 0.5 m/s. Which of the following vectors best represents the velocity of the canoe relative to shore?

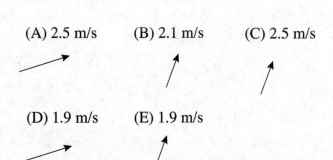

(A) 2.5 m/s   (B) 2.1 m/s   (C) 2.5 m/s

(D) 1.9 m/s   (E) 1.9 m/s

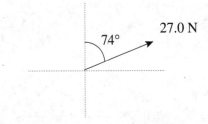

2. Force vector **A** has magnitude 27.0 N and is direction 74° from the vertical, as shown above. Which of the following are the horizontal and vertical components of vector **A**?

	Vertical	Horizontal
(A)	7.4 N	26.0 N
(B)	26.4 N	7.7 N
(C)	8.5 N	28.6 N
(D)	15.0 N	27.0 N
(E)	23.3 N	13.5 N

3. Which of the following is a scalar quantity?

(A) electric force
(B) gravitational force
(C) weight
(D) mass
(E) friction

› Solutions to Practice Problems

1. B—To solve, add the northward 2.0 m/s velocity vector to the eastward 0.5 m/s vector. These vectors are at right angles to one another, so the magnitude of the resultant is given by the Pythagorean theorem. You don't have a calculator on the multiple-choice section, though, so you'll have to be clever. There's only one answer that makes sense! The hypotenuse of a right triangle has to be bigger than either leg, but less than the algebraic sum of the legs. Only B, 2.1 m/s, meets this criterion.

2. A—Again, with no calculator, you cannot just plug into the calculator (though if you could, careful: the horizontal component of *A* is 27.0 N cos 16° because 16° is the angle from the horizontal.)

Answers B and E are wrong because the vertical component is bigger than the horizontal component, which doesn't make any sense based on the diagram. Choice C is wrong because the horizontal component is bigger than the magnitude of the vector itself—ridiculous! Same problem with choice D, where the horizontal component is equal to the magnitude of the vector. Answer must be A.

3. D—A scalar has no direction. All forces have direction, including weight (which is the force of gravity). Mass is just a measure of how much stuff is contained in an object, and thus has no direction.

CHAPTER 10

Free-Body Diagrams and Equilibrium

IN THIS CHAPTER

Summary: Free-body diagrams can help you see forces as vectors, and we'll review torque as well as a variety of forces: normal force, tension, friction, forces operating on inclined planes, and static and kinetic friction.

Key Ideas

✪ A free-body diagram is a picture that represents an object, along with the forces acting on that object.
✪ When the net force on an object equals zero, that object is in equilibrium.
✪ The normal force is *not* always equal to the weight of an object.
✪ Tension is a force applied by a rope or string.
✪ Friction is only found when there is contact between two surfaces.
✪ When an object is on an incline, use tilted axes, one parallel to the incline, one perpendicular.
✪ Torque occurs when a force is applied to an object, and that force can cause the object to rotate.

Relevant Equations

On an inclined plane, the weight vector can be broken into components:

$$mg_\perp = mg(\cos\theta)$$
$$mg_\parallel = mg(\sin\theta)$$

The force of friction is given by

$$F_f = \mu F_N$$

Physics, at its essence, is all about simplification. The universe is a complicated place, and if you want to make sense of it—which is what physicists try to do—you need to reduce it to some simplified representation: for example, with free-body diagrams.

We will refer regularly to forces. A force refers to a push or a pull applied to an object. Something can experience many different forces simultaneously—for example, you can push a block forward while friction pulls it backward, but the net force is the vector sum of all of the individual forces acting on the block.

> **Net Force:** The vector sum of all the forces acting on an object

What Is a Free-Body Diagram?

A free-body diagram is a picture that represents one or more objects, along with the forces acting on those objects. The objects are almost always drawn as rectangles or circles, just for the sake of simplicity, and the forces are always shown as vectors. Figure 10.1 shows a few examples.

Free-body diagrams are important because they help us to see forces as vectors. And if you can add vectors, you can analyze a free-body diagram. (If you can't add vectors, you didn't read Chapter 9 carefully enough.)

Let's look at the two examples in Figure 10.1. In the first, a force is directed down. This force, which is the force of gravity, was labeled in the diagram as "weight." The force of gravity on the hippo (that is, the hippo's weight) pulls downward. In the second example, a force is directed to the right. The pineapple is being pulled by a rope to the right.

> **Weight:** The force due to gravity, equal to the mass of an object times g, the gravitational field (about 10 N/kg on Earth)

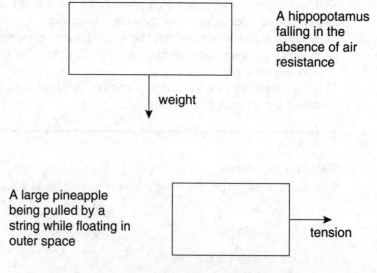

**Figure 10.1   Two examples of free-body diagrams.
As you see, there is no need to be artistic on the AP exam.**

You'll often see weight abbreviated as *mg*. Just be careful that the mass you use is in kilograms. For the rest of this chapter, we focus on objects in equilibrium.

Equilibrium

When the net force on an object equals zero, that object is in equilibrium. At equilibrium, an object is either at rest or *moving with a constant velocity*, but it is not accelerating.

You've heard of Newton's first law, of course: an object maintains its velocity unless acted upon by a net force. Well, an object in equilibrium is obeying Newton's first law.

How to Solve Equilibrium Problems

We have a tried-and-true method. Follow it every time you see an equilibrium situation.

1. Draw a proper free-body diagram.
2. Resolve force vectors into *x*- and *y*-components, if necessary.
3. Write an expression for the vector sum of the left–right forces. Then write an expression for the vector sum of the up–down forces. Set each of these expressions equal to zero.
4. Solve the resulting algebraic equations.

A Brief Interlude: UNITS!

Before we lose ourselves in the excitement of free-body diagrams, we need to pay tribute to the unit of force: the newton. One N (as newtons are abbreviated) equals one kg·m/s². We discuss why 1 newton equals 1 kg·m/s² in Chapter 12. For now, let it suffice that any force can be measured in newtons.

A Really Simple Equilibrium Problem

For those of you who prefer to splash your toes in the metaphorical swimming pool of physics before getting all the way in, this section is for you. Look at this situation:

Two astronauts tug on opposite sides of a satellite. The first astronaut tugs to the left with a force of 30 N. With what force does the second astronaut tug in order to keep the satellite at rest?

The solution to this problem is painfully obvious, but we'll go through the steps just to be thorough.

Step 1: Draw a proper free-body diagram.

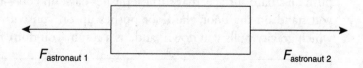

We can skip Step 2 because these vectors already line up with each other, so they do not need to be resolved into components.

Step 3: Write expressions for the vector sums.

This problem only involves left–right forces, so we only need one expression. Because we have an equilibrium situation, the net force is ZERO:

$$F_{net,x} = F_2 - F_1 = 0$$
$$F_2 - (30 \text{ N}) = 0$$

Step 4: Solve.

$$F_2 = 30 \text{ N}$$

Very good. Now, let's see how closely you were paying attention. Here's the same problem, with a slightly different twist.

> Two astronauts tug on opposite sides of a satellite. The first astronaut tugs to the left with a force of 30 N. With what force does the second astronaut tug in order to keep the satellite moving toward him at a constant speed of 20 m/s?

Think for a moment. Does the second astronaut have to apply more, less, or the same force as compared to the previous problem?

The second astronaut applies *exactly the same force* as in the previous problem! An object moving with constant velocity is in equilibrium, just as if the object were still. This is a central concept in Newtonian mechanics.

Normal Force

Let's return to Earth for a moment.

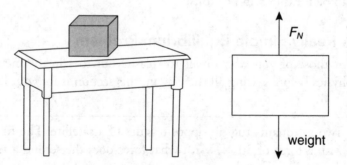

Figure 10.2   Normal force.

In Figure 10.2, a box is sitting on a table. The force of gravity pulls downward (as with the hippo, we've labeled this force "weight"). We know from experience that boxes sitting on tables do not accelerate downward; they remain where they are. Some force must oppose the downward pull of gravity.

This force is called the normal force,[1] and it is abbreviated F_N. In fact, whenever you push on a hard surface, that surface pushes back on you—it exerts a normal force. So, when you stand on the floor, the floor pushes up on you with the same amount of force with which gravity pulls you down, and, as a result, you don't fall through the floor.

> **Normal Force:** A force that acts perpendicular to the surface on which an object rests

---

[1]When physicists say "normal," they mean "perpendicular." The word "normal" in its conventional meaning simply does not apply to physicists.

 The normal force is *not* always equal to the weight of an object! Think about this before we get to the practice problems.

Tension

Tension is a force applied by a rope or string. Here are two of our favorite tension problems. The first is super easy, but a good introduction to tension; the second is more involved.

A box has a mass of 5 kg and is hung from the ceiling by a rope. What is the tension in the rope?

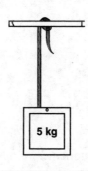

Step 1: Free-body diagram.

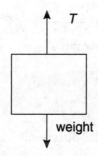

Step 2: Vector components.
Hey! These vectors already line up. On to Step 3.

Step 3: Equations.
Remember, weight is equal to mass times the gravitational field, or *mg*.

$$F_{net\ y} = T - mg = 0$$
$$T - (5\ \text{kg})\ (10\ \text{N/kg}) = 0$$

Step 4: Solve.

$$T = 50\ \text{N}$$

The same box is now hung by two ropes. One makes a 45-degree angle with the ceiling; the other makes a 30-degree angle with the ceiling. What is the tension in each rope?

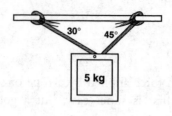

Step 1: Free-body diagram.

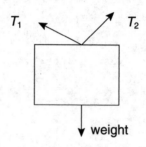

Step 2: Vector components.

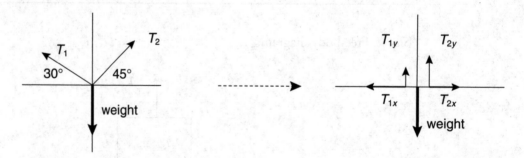

Step 3: Equations.
Let's start with the *x*-direction.

$$F_{net,x} = T_{1x} - T_{2x} = 0$$

And from vector analysis we know that

$$T_{1x} = T_1(\cos 30°) \text{ and } T_{2x} = T_2(\cos 45°)$$

so,

$$T_1(\cos 30°) - T_2(\cos 45°) = 0 \qquad \text{◁ } \textit{x-axis equation (Eq. 1)}$$

Similarly, if we look at the *y*-direction,

$$F_{net,y} = (T_{1y} + T_{2y}) - mg = 0$$

and

$$T_{1y} = T_1(\sin 30°) \text{ and } T_{2y} = T_2(\sin 45°)$$

so,

$$T_1(\sin 30°) + T_2(\sin 45°) - (50 \text{ N}) = 0. \quad \overleftarrow{\boxed{y\text{-axis equation (Eq. 2)}}}$$

Step 4: Solve.

We can solve Equation 1 and Equation 2 simultaneously and find T_1 and T_2. We'll let you do this on your own,[2] but in case you want to check your answers, $T_1 = 37$ N and $T_2 = 45$ N. (These are reasonable answers, as the tension in each rope is the same power of 10 as the 50 N weight of the box.)

Steps 1, 2, and 3 are the important steps. Step 4 only involves math. "ONLY math?!?" you ask, incredulous. That's the toughest part!

Well, maybe for some people. Getting the actual correct answer does depend on your algebra skills. But, and this is important, *this is AP Physics, NOT AP Algebra.* The graders of the AP exam will assign most of the credit just for setting up the problem correctly! If you're stuck on the algebra, skip it! Come up with a reasonable answer for the tensions, and move on!

We're not kidding. Look at Chapter 7, which discusses approaches to the free-response section, for more about the relative importance of algebra.

Friction

Friction is only found when there is contact between two surfaces.

> **Friction:** A force acting parallel to two surfaces in contact. If an object moves, the friction force always acts opposite the direction of motion.

For example, let's say you slide a book at a constant speed across a table. The book is in contact with the table, and, assuming your table isn't frictionless, the table will exert a friction force on the book opposite its direction of motion. Figure 10.3 shows a free-body diagram of that situation.

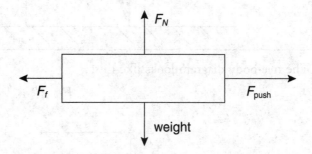

Figure 10.3 Free-body diagram of a book sliding on a table.

---

[2]Try solving the *x*-axis equation for T_1, then plug that into the *y*-axis equation:

$$T_1 = T_2 \frac{\cos 45°}{\cos 30°}$$

$$\left(T_2 \frac{\cos 45°}{\cos 30°}\right) \sin 30° + (T_2 \sin 45°) - (50 \text{ N}) = 0$$

Plug in the value of $\cos 45°$, $\cos 30°$, $\sin 30°$, $\sin 45°$ . . . and now it's easy to solve for $T_2 = 45$ N.

We know that because the book represented in Figure 10.3 is not being shoved through the table or flying off it, F_N must equal the book's weight. And because the book moves at constant velocity, the force you exert by pushing the book, F_{push}, equals the force of friction, F_f. *Remember, being in equilibrium does not necessarily mean that the book is at rest. It could be moving at a constant velocity.*

How do we find the magnitude of F_f?

$$F_f = \mu \, F_N$$

Mu (μ) is the coefficient of friction. This is a dimensionless number (that is, it doesn't have any units) that describes how big the force of friction is between two objects. It is found experimentally because it differs for every combination of materials (for example, if a wood block slides on a glass surface), but it will usually be given in AP problems that involve friction.

And if μ isn't given, it is easy enough to solve for—just rearrange the equation for μ algebraically:

$$\mu = \frac{F_f}{F_N}.$$

Remember, when solving for F_f, do not assume that F_N equals the weight of the object in question. Here's a problem where this reminder comes in handy:

A floor buffer consists of a heavy base (m = 15 kg) attached to a very light handle. A worker pushes the buffer by exerting a force P directly down the length of the handle. If the coefficient of friction between the buffer and the floor is μ = 0.36, what is the magnitude of the force P needed to keep the buffer moving at a constant velocity?

The free-body diagram looks like this:

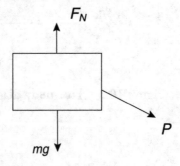

Exam tip from an AP Physics veteran:
When drawing a free-body diagram, put the tail of the force vectors on the object, with the arrow pointing away from the object. Never draw a force vector pointing into an object, even when something is pushing, as with the P force in this example.

—*Chris, high school junior*

Now, in the vertical direction, there are three forces acting: F_N acts up; weight and the vertical component of P act down.

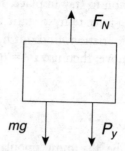

Notice that when we set up the equilibrium equation in the vertical direction, $F_N - (mg + P_y) = 0$, we find that F_N is greater than mg.

Let's finish solving this problem together. We've already drawn the vertical forces acting on the buffer, so we just need to add the horizontal forces to get a complete free-body diagram with the forces broken up into their components (Steps 1 and 2):

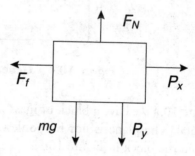

Step 3 calls for us to write equations for the vertical and horizontal directions. We already found the equilibrium equation for the vertical forces,

$$F_N - (mg + P_y) = 0$$

and it's easy enough to find the equation for the horizontal forces,

$$F_f - P_x = 0$$

To solve this system of equations (Step 4), we can reduce the number of variables with a few substitutions. For example, we can rewrite the equation for the horizontal forces as

$$\mu \cdot F_N - P \cdot \cos 37° = 0$$

Furthermore, we can use the vertical equation to substitute for F_N,

$$F_N = mg + P \cdot \sin 37°$$

Plugging this expression for F_N into the rewritten equation for the horizontal forces, and then replacing the variables m, g, and μ with their numerical values, we can solve for P. The answer is $P = 93$ N.

Static and Kinetic Friction

You may have learned that the coefficient of friction takes two forms: **static** and **kinetic** friction. Use the coefficient of static friction if something is stationary, and the coefficient of kinetic friction if the object is moving. The equation for the force of friction is essentially the same in either case: $F_f = \mu F_N$.

The only strange part about static friction is that the coefficient of static friction is a *maximum* value. Think about this for a moment . . . if a book just sits on a table, it doesn't need any friction to stay in place. But that book won't slide if you apply a very small horizontal pushing force to it, so static friction can act on the book. To find the maximum coefficient of static friction, find out how much horizontal pushing force will just barely cause the book to move; then use $F_f = \mu F_N$.

Inclined Planes

These could be the most popular physics problems around. You've probably seen way too many of these already in your physics class, so we'll just give you a few tips on approaching them.

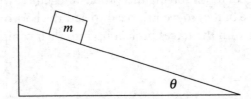

Figure 10.4   Generic inclined-plane situation.

In Figure 10.4 we have a block of mass m resting on a plane elevated an angle θ above the horizontal. The plane is not frictionless. We've drawn a free-body diagram of the forces acting on the block in Figure 10.5a.

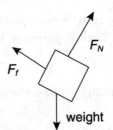

Figure 10.5a   Forces acting on the block in Figure 10.4.

F_f is directed parallel to the surface of the plane, and F_N is, by definition, directed perpendicular to the plane. It would be a pain to break these two forces into x- and y-components, so instead we will break the "weight" vector into components that "line up" with F_f and F_N, as shown in Figure 10.5b.

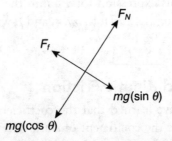

Figure 10.5b   Forces acting on the block in Figure 10.4, with the weight vector resolved into components that line up with the friction force and the normal force.

Memorize this

> *As a rule of thumb, in virtually all inclined-plane problems, you can always break the weight vector into components parallel and perpendicular to the plane, WHERE THE COMPONENT PARALLEL TO (POINTING DOWN) THE PLANE = mg(sin θ) AND THE COMPONENT PERPENDICULAR TO THE PLANE = mg(cos θ).*

This rule always works, as long as the angle of the plane is measured from the horizontal.

Torque

Torque occurs when a force is applied to an object, and that force can cause the object to rotate.

$$\textbf{Torque} = Fd$$

In other words, the torque exerted on an object equals the force exerted on that object (F) multiplied by the distance between where the force is applied and the fulcrum (d) as long as the force acts perpendicular to the object.

Fulcrum: The point about which an object rotates

Figure 10.6 shows what we mean:

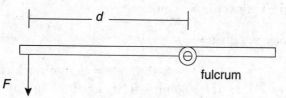

Figure 10.6   The torque applied to this bar equals *Fd.*

The unit of torque is the newton-meter.

Torque problems on the AP Physics B exam tend to be pretty straightforward. To solve them, set counterclockwise torques equal to clockwise torques.

Here's an example.

Bob is standing on a bridge. The bridge itself weighs 10,000 N. The span between pillars *A* and *B* is 80 m. Bob is 20 m from the center of the bridge. His mass is 100 kg. Assuming that the bridge is in equilibrium, find the force exerted by pillar *B* on the bridge.

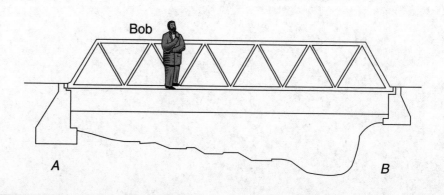

Step 1: Free-body diagram.

We'll use point A as the fulcrum to start with. Why? In a static equilibrium situation, since the bridge isn't *actually* rotating, any point on the bridge could serve as a fulcrum. But we have two unknown forces here, the forces of the supports A and B. We choose the location of one of these supports as the fulcrum, because now that support provides zero torque—the distance from the fulcrum becomes zero! Now all we have to do is solve for the force of support B.

The diagram below isn't a true "free-body diagram," because it includes both distance and forces, but it is useful for a torque problem. Bob's weight acts downward right where he stands.

The bridge's weight is taken into account with a force vector acting at the bridge's center of mass; that is, 40 m to the right of pillar A. This is a generally valid approach—replace the weight of an extended object with a single weight vector acting at the center of mass.

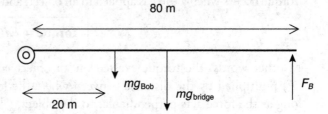

Step 2: Vector components.

We don't have to worry about vector components here. (We would have if the forces had not acted perpendicular to the bridge. Such a situation is much more likely to appear on the C than the B exam.)

Step 3: Equations.

$$\text{Torque}_{net} = \text{counterclockwise} - \text{clockwise} = 0$$
$$(F_B)(80 \text{ m}) - [(100 \text{ kg} \cdot 10 \text{ N/kg})(20 \text{ m}) + (10{,}000 \text{ N})(40 \text{ m})] = 0$$

Step 4: Solve. $F_B = 5300$ N

This is reasonable because pillar B is supporting *less* than half of the 11,000 N weight of the bridge and Bob. Because Bob is closer to pillar A, and otherwise the bridge is symmetric, A should bear the majority of the weight.

› Practice Problems

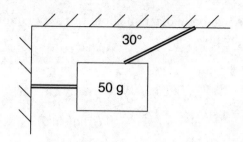

1. A 50-g mass is hung by string as shown in the picture above. The left-hand string is horizontal; the angled string measures 30° to the horizontal. What is the tension in the angled string?

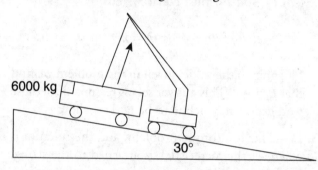

2. A 6000-kg bus sits on a 30° incline. A crane attempts to lift the bus off of the plane. The crane pulls perpendicular to the plane, as shown in the diagram. How much force must the crane apply so that the bus is suspended just above the surface? [cos 30° = 0.87, sin 30° = 0.50]

(A) 52,000 N
(B) 30,000 N
(C) 6000 N
(D) 5200 N
(E) 300 N

3. Give two examples of a situation in which the normal force on an object is less than the object's weight. Then give an example of a situation in which there is NO normal force on an object.

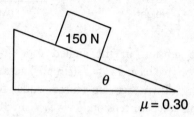

4. A 150-N box sits motionless on an inclined plane, as shown above. What is the angle of the incline?

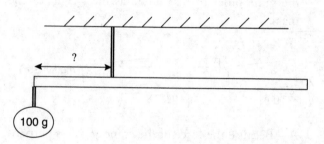

5. A 50-g meterstick is to be suspended by a single string. A 100-g ball hangs from the left-hand edge of the meterstick. Where should the string be attached so that the meterstick hangs in equilibrium?

(A) at the left-hand edge
(B) 40 cm from left-hand edge
(C) 30 cm from right-hand edge
(D) 17 cm from left-hand edge
(E) at the midpoint of the meterstick

› Solutions to Practice Problems

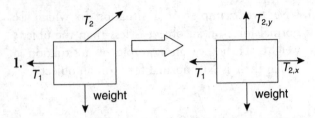

1.

Call the tension in the angled rope T_2. In the y-direction, we have $T_{2,y} = T_2(\sin 30°)$ acting up, and mg acting down. Set "up" forces equal to "down" forces and solve for tension: $T_2 = mg/(\sin 30°)$. Don't forget to use the mass in KILOgrams, i.e., 0.050 kg. The tension thus is $(0.050 \text{ kg})(10 \text{ N/kg})/(0.5) = 1.0$ N. This is reasonable because the tension is about the same order of magnitude as the weight of the mass.

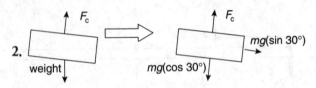

2.

A—Because the force of the crane, F_c, acts perpendicular to the plane, the parallel-to-the-plane direction is irrelevant. So all we need to do is set F_c equal to $mg(\cos 30°) = (6000 \text{ kg})(10 \text{ N/kg})(.87)$ and plug in. $F_c = 52,000$ N. This is a reasonable answer because it is less than—but on the same order of magnitude as—the weight of the bus.

3. When a block rests on an inclined plane, the normal force on the block is less than the block's weight, as discussed in the answer to #2. Another example in which the normal force is less than an object's weight occurs when you pull a toy wagon.

In any situation where an object does not rest on a surface (for example, when something floats in space), there is no normal force.

4. This free-body diagram should be very familiar to you by now.

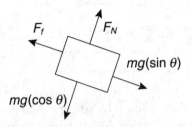

The box is in equilibrium, so F_f must equal $mg(\sin \theta)$, and F_N must equal $mg(\cos \theta)$.

$$\mu \cdot F_N = \mu \cdot mg(\cos \theta) = mg(\sin \theta).$$

Plugging in the values given in the problem we find that $\mu = 17°$. This answer seems reasonable because we'd expect the incline to be fairly shallow.

5. D—This is a torque problem, and the fulcrum is wherever the meter stick is attached to the string. We know that the meter stick's center of mass is at the 50 cm mark, so we can draw the following picture.

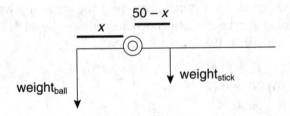

Because the stick is in equilibrium, the clockwise torques equal the counterclockwise torques: $(1 \text{ N})(x) = (0.5 \text{ N})(50 - x)$. So $x =$ something in the neighborhood of $25/1.5 \sim 17$ cm. This answer is less than 50 cm, and is closer to the edge with the heavy mass, so it makes sense.

〉 Rapid Review

- A free-body diagram is a simplified representation of an object and the forces acting on it.

- When the net force on an object is zero, it is in equilibrium. This means that it is either at rest or that it is moving at a constant velocity.

- To solve an equilibrium problem, draw a good free-body diagram, resolve all forces into *x*- and *y*-components, and then set the vector sum of the *x*-components equal to zero and the vector sum of the *y*-components equal to zero.

- The units of force are newtons, where $1 \text{ N} = 1 \text{ kg·m/s}^2$.

- Torque equals the force exerted on an object multiplied by the distance between where that force is applied and the fulcrum (the point about which an object can rotate). When an object is in equilibrium, the counterclockwise torques equal the clockwise torques.

- A "normal force" means the force of a solid surface pushing perpendicular to that surface. The normal force is NOT always equal to an object's weight.

CHAPTER 11

Kinematics

IN THIS CHAPTER

Summary: As soon as an object's velocity changes, you need to analyze the problem using kinematics, which deals with aspects of motion separate from considerations of mass and force.

Key Ideas

- ✪ Kinematics problems involve five variables: initial velocity, final velocity, displacement, acceleration, and time interval.
- ✪ Use the three kinematics equations whenever acceleration is constant.
- ✪ Average speed is the total distance in a given time divided by the time it takes you to travel that distance.
- ✪ Velocity is just like speed, except it's a vector.
- ✪ Acceleration is the change in velocity divided by a time interval.
- ✪ Displacement is the vector equivalent of distance.
- ✪ The key rule of projectile motion is that an object's motion in one dimension does not affect its motion in any other dimension.

Relevant Equations

The constant-acceleration kinematics equations, which we refer to as the "star" equations:

$$* \ v_f = v_0 + at$$

$$** \ \Delta x = v_0 t + \tfrac{1}{2}at^2$$

$$*** \ v_f^2 = v_0^2 + 2a\Delta x$$

The equilibrium problems we saw in the last chapter all had something in common: there was no acceleration. Sure, an object can move at a constant velocity and still be in equilibrium, but as soon as an object's velocity changes, you need a new set of tricks to analyze the situation. This is where kinematics comes in.

Velocity, Acceleration, and Displacement

We'll start with a few definitions.

> **Average Speed:** $\dfrac{\Delta x}{\Delta t}$ The units for speed are m/s.

In this definition, Δx means "displacement" and Δt means "time interval." Average speed is the total displacement you travel in a straight line in a given time divided by the time it takes you to travel that distance. This is different from "instantaneous speed," which is your speed at any given moment. WARNING: The formula you learned in seventh grade, "speed = distance/time" is ONLY valid for an average speed, or when something is moving with constant speed. If an object speeds up or slows down, and you want to know its speed at some specific moment, don't use this equation![1]

> **Velocity:** Just like speed, except it's a vector

Questions on the AP exam tend to focus on velocity more than speed, because velocity says more about an object's motion. (Remember, velocity has both magnitude and direction.) Acceleration occurs when an object changes velocity.

> **Acceleration:** $\dfrac{\Delta v}{\Delta t}$ The units of acceleration are m/s^2.

The symbol Δ means "change in." So $\Delta v = v_f - v_0$, where v_f means "final velocity" and v_0 means "initial velocity" and is pronounced "v-naught." Similarly, Δt is the time interval during which this change in velocity occurred.

Just as velocity is the vector equivalent of speed, displacement is the vector equivalent of distance—it has both magnitude and direction.

> **Displacement:** The vector equivalent of distance

So, let's say that you head out your front door and walk 20 m south. If we define north to be the positive direction, then your displacement was "−20 m." If we had defined south to be the positive direction, your displacement would have been "+20 m." Regardless of which direction was positive, the *distance* you traveled was just "20 m." (Or consider this: If you walk 20 m north, followed by 5 m back south, your displacement is 15 m, north. Your displacement is *not* 25 m.)

---

[1]Use the "star equations," which we will address in detail momentarily.

Fundamental Kinematics Equations

Putting all of these definitions together, we can come up with some important lists. First, we have our five variables:

> **Variables**
> 1. v_0 (initial velocity)
> 2. v_f (final velocity)
> 3. Δx (displacement)
> 4. a (acceleration)
> 5. t (time interval)

Using just these five variables, we can write the three most important kinematics equations. An important note: *The following equations are valid ONLY when acceleration is constant. We repeat: ONLY WHEN ACCELERATION IS CONSTANT.* (Which is most of the time, in physics B.)[2]

> **Equations**
> * $v_f = v_0 + at$
> ** $\Delta x = v_0 t + \frac{1}{2} at^2$
> *** $v_f^2 = v_0^2 + 2a\Delta x$

We call these equations the "star equations." You don't need to call them the "star equations," but just be aware that we'll refer to the first equation as "*," the second as "**," and the third as "***" throughout this chapter.

These are the only equations you really need to memorize for kinematics problems.

Kinematics Problem-Solving

Step 1: Write out all five variables in a table. Fill in the known values, and put a "?" next to the unknown values.

Step 2: Count how many known values you have. If you have three or more, move on to Step 3. If you don't, find another way to solve the problem (or to get another known variable).

Step 3: Choose the "star equation" that contains all three of your known variables. Plug in the known values, and solve.

Step 4: Glory in your mastery of physics. Feel proud. *Put correct units on your answer.*

Be sure that you have committed these steps to memory. Now, let's put them into action.

> A rocket-propelled car begins at rest and accelerates at a constant rate up to a velocity of 120 m/s. If it takes 6 s for the car to accelerate from rest to 60 m/s, how long does it take for the car to reach 120 m/s, and how far does it travel in total?

---

[2]When *can't* you use kinematics, you ask? The most common situations are when a mass is attached to a spring, when a roller coaster travels on a curvy track, or when a charge is moving in an electric field produced by other charges. To approach these problems, use conservation of energy, as discussed in Chapter 14.

Before we solve this problem—or any problem, for that matter—we should think about the information it provides. The problem states that acceleration is constant, so that means we can use our kinematics equations. Also, it asks us to find two values, a time and a distance. Based on the information in the problem, we know that the time needed for the car to reach 120 m/s is greater than 6 s because it took 6 s for the car just to reach 60 m/s. Moreover, we can estimate that the car will travel several hundred meters in total, because the car's *average* velocity must be less than 120 m/s, and it travels for several seconds.

So now let's solve the problem. We'll use our four-step method.

Step 1: Table of variables.
The car begins at rest, so $v_0 = 0$ m/s. The final velocity of the car is 120 m/s. We're solving for time and displacement, so those two variables are unknown. And, at least for right now, we don't know what the acceleration is.

v_0	0 m/s
v_f	120 m/s
Δx	?
a	?
t	?

This table represents the car's entire motion.

Step 2: Count variables.
We only have two values in our chart, but we need three values in order to use our kinematics equations. Fortunately, there's enough information in the problem for us to solve for the car's acceleration.

Acceleration is defined as a change in velocity divided by the time interval during which that change occurred. The problem states that in the first 6 s, the velocity went from 0 m/s to 60 m/s.

$$\Delta v = v_f - v_0 = 60 \text{ m/s} - 0 = 60 \text{ m/s}$$

$$a = \frac{\Delta v}{\Delta t} = \frac{60 \text{ m/s}}{6\text{s}} = 10 \text{ m/s}^2$$

v_0	0 m/s
v_f	120 m/s
Δx	?
a	10 m/s^2
t	?

Chart for car's entire motion.

We now have values for three of our variables, so we can move to Step 3.

Step 3: Use "star equations" to solve.
All three of our known values can be plugged into *, which will allow us to solve for *t*.

$$* \ v_f = v_0 + at$$
$$120 \text{ m/s} = 0 + (10 \text{ m/s}^2)t$$
$$t = 12 \text{ s}$$

Now that we know t, we can use either ** or *** to solve for displacement. Let's use **:

$$** \Delta x = v_0 t + \tfrac{1}{2} a t^2$$
$$\Delta x = 0(12 \text{ s}) + \tfrac{1}{2}(10 \text{ m/s}^2)(12 \text{ s})^2$$
$$\Delta x = 720 \text{ m}$$

Step 4: Units.

Always remember units! And make sure that your units are sensible—if you find that an object travels a distance of 8 m/s, you've done something screwy. In our case, the answers we found have sensible units. Also, our answers seem reasonable based on the initial estimates we made: It makes sense that the car should travel a bit more than 6 s, and it makes sense that it should go several hundred meters (about half a mile) in that time.

Free Fall

Problems that involve something being thrown off a cliff[3] are great, because vertical acceleration in these problems equals g in just about every case.

> g: The acceleration due to gravity near the Earth's surface; about 10 m/s^2

Falling-object problems should be solved using the method we outlined above. However, you have to be really careful about choosing a positive direction and sticking to it. That is, figure out before you solve the problem whether you want "up" to be positive (in which case a equals -10 m/s^2) or "down" to be positive (where a would therefore equal $+10$ m/s^2).

> **Exam tip from an AP Physics veteran:**
> You may remember that a more precise value for g is 9.80 m/s^2. That's correct. But estimating g as 10 m/s^2 is encouraged by the AP readers to make calculation quicker.
>
> —*Jake, high school junior*

Here's a practice problem:

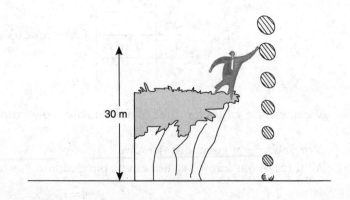

You are standing on a cliff, 30 m above the valley floor. You throw a watermelon vertically upward at a velocity of 3 m/s. How long does it take until the watermelon hits the valley floor?

30 m

---

[3]The writers of the AP exam *love* to throw things off cliffs.

Begin by defining the positive direction. We will call "up" positive. Then use the four-step method to solve the problem.

Step 1: Table of variables.

v_0	3 m/s
v_f	?
Δx	−30 m
a	−10 m/s^2
t	?

Melon's entire motion

Why do we always indicate what part of the motion the kinematics chart is for? Well, this problem *could* be solved instead by making two separate charts: one for the upward motion (where v_f would be zero), and one for the downward motion (where v_o would be zero). Be crystal clear how much of an object's motion you are considering with a chart.

Remember that displacement is a vector quantity. Even though the melon goes up before coming back down, the displacement is simply equal to the height at which the melon ends its journey (0 m) minus its initial height (30 m). Another way to think about displacement: In total, the melon ended up 30 m BELOW where it started. Because down is the negative direction, the displacement is −30 m.

Step 2: Count variables.
Three! We can solve the problem.

Step 3: Solve.
The rest of this problem is just algebra. Yes, you have to do it right, but setting up the problem correctly and coming up with an answer that's reasonable is more important than getting the exact right answer. Really! If this part of an AP free-response problem is worth 5 points, you might earn 4 of those points just for setting up the equation and plugging in values correctly, even if your final answer is wrong.

But which equation do you use? We have enough information to use ** ($x - x_0 = v_0t + \frac{1}{2}at^2$) to solve for t. Note that using ** means that we'll have to solve a quadratic equation; you can do this with the help of the quadratic formula.[4] Or, if you have a graphing calculator and you're working on the free-response portion of the exam (where calculators are allowed), you can use your calculator to solve the equation.

$$** \; x - x_0 = v_0t + \frac{1}{2}at^2$$
$$(-30 \text{ m}) = (3 \text{ m/s})t + \frac{1}{2}(-10 \text{ m/s}^2)t^2$$
$$t = 2.8 \text{ s}$$

Algebra hint: You can avoid quadratics in all Physics B kinematics problems by solving in a roundabout way. Try solving for the velocity when the watermelon hits the ground using *** [$v_f^2 = v_0^2 + 2a(x - x_0)$]; then plug into * ($v_f = v_0 + at$). This gives you the same answer.

---

[4]
$$x = \frac{-b \pm \sqrt{b^2 - 4a \cdot c}}{2a}$$

Projectile Motion

Things don't always move in a straight line. When an object moves in two dimensions, we look at vector components.

The super-duper-important general rule is this: *An object's motion in one dimension does not affect its motion in any other dimension.*

The most common kind of two-dimensional motion you will encounter is projectile motion. The typical form of projectile-motion problems is the following:

"A ball is shot at a velocity v *from a cannon pointed at an angle* θ *above the horizontal . . ."*

No matter what the problem looks like, remember these rules:

- The vertical component of velocity, v_y, equals $v(\sin \theta)$.
- The horizontal component of velocity, v_x, equals $v(\cos \theta)$ when θ is measured relative to the horizontal.
- Horizontal velocity is constant.
- Vertical acceleration is g, directed downward.

Here's a problem that combines all of these rules:

A ball is shot at a velocity 25 m/s from a cannon pointed at an angle $\theta = 30°$ above the horizontal. How far does it travel before hitting the level ground?

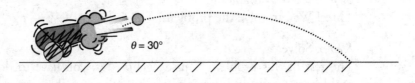

θ = 30°

We begin by defining "up" to be positive and writing our tables of variables, one for horizontal motion and one for vertical motion.

	Horizontal		Vertical	
v_0	$v(\cos \theta) = 22\,\text{m/s}$	v_0	$v(\sin \theta) = 13\,\text{m/s}$	
v_f	$v(\cos \theta) = 22\,\text{m/s}$	v_f	?	
Δx	?	Δx	0 m	Entire motion of cannon ball
a	$0\,\text{m/s}^2$	a	$-10\,\text{m/s}^2$	
t	?	t	?	

Note that because horizontal velocity is constant, on the horizontal table, $v_f = v_0$, and $a = 0$. Also, because the ball lands at essentially the same height it was launched from, $\Delta x = 0$ on the vertical table. You should notice, too, that we rounded values in the tables to two significant figures (for example, we said that v_0 in the vertical table equals 13 m/s, instead of 12.5 m/s). We can do this because the problem is stated using only two significant figures for all values, so rounding to two digits is acceptable, and it makes doing the math easier for us.

We know that t is the same in both tables—the ball stops moving horizontally at the same time that it stops moving vertically (when it hits the ground). We have enough information in the vertical table to solve for t by using equation **.

$$** \, x - x_0 = v_0 t + \tfrac{1}{2} a t^2$$
$$0 = (13 \text{ m/s})t + \tfrac{1}{2}(-10 \text{ m/s}^2)t^2$$
$$t = 2.6 \text{ s}$$

Using this value for t, we can solve for $x - x_0$ in the horizontal direction, again using **.

$$** \, x - x_0 = v_0 t + \tfrac{1}{2} a t^2$$
$$x - x_0 = (22 \text{ m/s})(2.6 \text{ s}) + 0$$
$$x - x_0 = 57 \text{ m}$$

The cannonball traveled 57 m, about half the length of a football field.

You may have learned in your physics class that the range of a projectile (which is what we just solved for) is

$$R = \frac{v^2 \sin 2\theta}{g}$$

If you feel up to it, you can plug into this equation and show that you get the same answer we just got. There's no need to memorize the range equation, but it's good to know the conceptual consequences of it: The range of a projectile on *level earth* depends only on the initial speed and angle, and the maximum range is when the angle is 45°.

A Final Word About Kinematics Charts

The more you practice kinematics problems using our table method, the better you'll get at it, and the quicker you'll be able to solve these problems. Speed is important on the AP exam, and you can only gain speed through practice. So use this method on all your homework problems, and when you feel comfortable with it, you might want to use it on quizzes and tests. The other benefit to the table method, besides speed, is consistency; it forces you to set up every kinematics problem the same way, every time. This is a time-tested strategy for success on the AP exam.

Motion Graphs

You may see some graphs that relate to kinematics on the AP test. They often look like those in Figure 11.1:

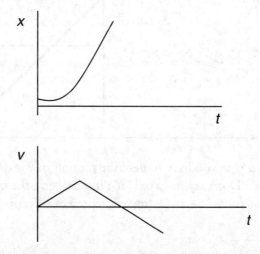

Figure 11.1 Typical motion graphs. (As an exercise, you may want to describe the motion these represent; answers are at the end of the chapter.)

We call these graphs by the names of their axes: For example, the top graph in Figure 11.1 is a "position–time graph" and the second one is a "velocity–time graph."

Here are some rules to live by:

- The slope of a position–time graph at any point is the velocity of the object at that point in time.
- The slope of a velocity–time graph at any point is the acceleration of the object at that point in time.
- The area under a velocity–time graph between two times is the displacement of the object during that time interval.

It's sometimes confusing what is meant by the area "under" a graph. In the velocity–time graphs below, the velocity takes on both positive and negative values. To find the object's displacement, we first find the area above the *t*-axis; this is positive displacement. Then we subtract the area below the *t*-axis, which represents negative displacement. The correct area to measure is shown graphically in Figure 11.2a. Whatever you do, *don't* find the area as shown in Figure 11.2b! When we say "area," we measure that area to the *t*-axis only.

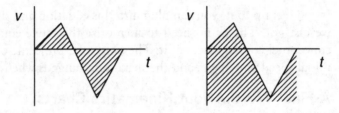

Figure 11.2a: Do this.　　　　**Figure 11.2b: Don't do this.**

Problems involving graphical analysis can be tricky because they require you to think abstractly about an object's motion. For practice, let's consider one of the most common velocity–time graphs you'll see:

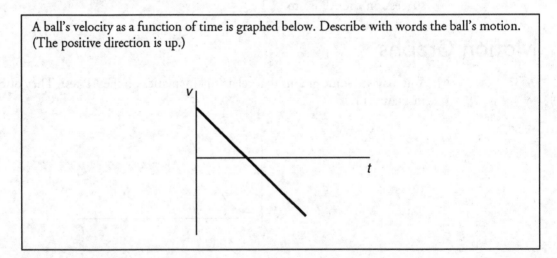

A ball's velocity as a function of time is graphed below. Describe with words the ball's motion. (The positive direction is up.)

Whenever you have to describe motion in words, do so in everyday language, not physics-speak. Don't say the word "it"; instead, give the object some specificity. Never say "positive" or "negative"; + and − merely represent directions, so name these directions.[5]

---

[5]Why *shouldn't* I say "positive" and "negative," you ask? Well, how do these directions to the store sound: "Define north as positive. Start from zero, and go positive 20 constantly; then come back at −20, also constantly." You'd never say that! But, this is what you'll sound like unless you use common language.

In this case, let's consider a ball going up (positive) and down (negative). Here's how we'd answer the question:

"At first the ball is moving upward pretty fast, but the ball is slowing down while going upward. (I know this because the speed is getting closer to zero in the first part of the graph.) The ball stops for an instant (because the v–t graph crosses the horizontal axis); then the ball begins to speed up again, but this time moving downward."

Now, no numerical values were given in the graph. But would you care to hazard a guess as to the likely slope of the graph's line if values *were* given?[6]

Figure 11.1 Graphs

The position–time graph has a changing slope, so the speed of the object is changing. The object starts moving one way, then stops briefly (where the graph reaches its minimum, the slope, and thus the speed is zero). The object then speeds up in the other direction. How did I know the object's velocity changed direction? The position was at first *approaching* the origin, but then was getting farther away from the origin.

The second graph is a velocity–time graph, it must be analyzed differently. The object starts from rest, but speeds up; the second part of the motion is just like the example shown above, in which the object slows to a brief stop, turns around, and speeds up.

---

[6] $-10 m/s^2$, if we're on Earth.

› Practice Problems

Multiple Choice:

1. A firework is shot straight up in the air with an initial speed of 50 m/s. What is the maximum height it reaches?

   (A) 12.5 m
   (B) 25 m
   (C) 125 m
   (D) 250 m
   (E) 1250 m

2. On a strange, airless planet, a ball is thrown downward from a height of 17 m. The ball initially travels at 15 m/s. If the ball hits the ground in 1 s, what is this planet's gravitational acceleration?

   (A) 2 m/s²
   (B) 4 m/s²
   (C) 6 m/s²
   (D) 8 m/s²
   (E) 10 m/s²

Free Response:

3. An airplane attempts to drop a bomb on a target. When the bomb is released, the plane is flying upward at an angle of 30° above the horizontal at a speed of 200 m/s, as shown below. At the point of release, the plane's altitude is 2.0 km. The bomb hits the target.

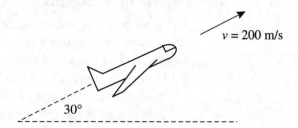

   (a) Determine the magnitude and direction of the vertical component of the bomb's velocity at the point of release.
   (b) Determine the magnitude and direction of the horizontal component of the bomb's velocity at the point when the bomb contacts the target.
   (c) Determine how much time it takes for the bomb to hit the target after it is released.
   (d) At the point of release, what angle below the horizontal does the pilot have to look in order to see the target?

› Solutions to Practice Problems

1. Call "up" the positive direction, and set up a chart. We know that $v_f = 0$ because, at its maximum height, the firework stops for an instant.

v_0	+50 m/s
v_f	0
Δx	?
a	−10 m/s^2
t	?

Solve for Δx using equation ***: $v_f^2 = v_0^2 + 2a(\Delta x)$. Answer is (C) 125 m, or about skyscraper height.

2. Call "down" positive, and set up a chart:

v_0	+15 m/s
v_f	?
Δx	+17 m
a	?
t	1 s

Plug straight into ** ($\Delta x = v_0 t + \frac{1}{2}at^2$) and you have the answer. This is NOT a quadratic, because this time t is a known quantity. The answer is (B) 4 m/s^2, less than half of Earth's gravitational field, but close to Mars's gravitational field.

3. (a) Because the angle 30° is measured to the horizontal, the magnitude of the vertical component of the velocity vector is just (200 m/s) (sin 30°), which is 100 m/s. The direction is "up," because the plane is flying up.

(b) The horizontal velocity of a projectile is constant. Thus, the horizontal velocity when the bomb hits the target is the same as the horizontal velocity at release, or (200 m/s)(cos 30°) = 170 m/s, to the right.

(c) Let's call "up" the positive direction. We can solve this projectile motion problem by our table method.

Horizontal			Vertical	
v_0	+170 m/s		v_0	+100 m/s
v_f	+170 m/s		v_f	?
Δx	?		Δx	−2000 m
a	0		a	−10 m/s^2
t	?		t	?

Don't forget to convert to meters, and be careful about directions in the vertical chart.

The horizontal chart cannot be solved for time; however, the vertical chart can. Though you could use the quadratic formula or your fancy calculator to solve $x - x_0 = v_0 t + \frac{1}{2}at^2$, it's much easier to start with ***, $v_f^2 = v_0^2 + 2a(x - x_0)$, to find that v_f vertically is −220 m/s (this velocity must have a negative sign because the bomb is moving down when it hits the ground). Then, plug in to *($v_f = v_0 + at$) to find that the bomb took 32 s to hit the ground.

(d) Start by finding how far the bomb went horizontally. Because horizontal velocity is constant, we can say distance = velocity × time. Plugging in values from the table, distance = (170 m/s)(32 s) = 5400 m. Okay, now look at a triangle:

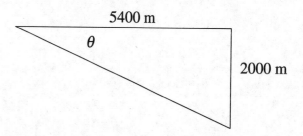

By geometry, $\tan\theta$ = 2000 m/5400 m. The pilot has to look down at an angle of 20°.

❯ Rapid Review

- Average speed is total distance divided by total time. Instantaneous speed is your speed at a particular moment.

- Velocity is the vector equivalent of speed.

- Acceleration is a change in velocity divided by the time during which that change occurred.

- Displacement is the vector equivalent of distance.

- The three "star equations" are valid only when acceleration is constant.

- To solve any kinematics problem, follow these four steps:
  - Write out a table containing all five variables—v_0, v_f, $x - x_0$, a, t—and fill in whatever values are known.
  - Count variables. If you have three known values, you can solve the problem.
  - Use the "star equation" that contains your known variables.
  - Check for correct units.

- When an object falls (in the absence of air resistance), it experiences an acceleration of g, about 10 m/s². It's particularly important for problems that involve falling objects to define a positive direction before solving the problem.

- An object's motion in one dimension does not affect its motion in any other dimension.

- Projectile motion problems are usually easier to solve if you break the object's motion into "horizontal" and "vertical" vector components.

- The slope of a distance–time graph is velocity.

- The slope of a velocity–time graph is acceleration.

- The area under a velocity–time graph is displacement.

CHAPTER 12

Newton's Second Law, $F_{net} = ma$

IN THIS CHAPTER

Summary: Chapter 10 explained how to deal with objects in equilibrium, that is, with zero acceleration. The same problem-solving process can be used with accelerating objects.

Key Ideas

- ✪ Only a NET force, not an individual force, can be set equal to *ma*.
- ✪ Use a free-body diagram and the four-step problem-solving process when a problem involves forces.
- ✪ When two masses are connected by a rope, the rope has the same tension throughout. (One rope = one tension.)
- ✪ Newton's third law: force pairs must act on different objects.

Relevant Equations

Um, the chapter title says it all . . .

$$F_{net} = ma$$

What this means is that *the net force acting on an object is equal to the mass of that object multiplied by the object's acceleration.* And that statement will help you with all sorts of problems.

The Four-Step Problem-Solving Process

If you decide that the best way to approach a problem is to use $F_{net} = ma$, then you should solve the problem by following these four steps.

1. Draw a proper free-body diagram.
2. Resolve vectors into their components.
3. For each axis, set up an expression for F_{net}, and set it equal to ma.
4. Solve your system of equations.

Note the marked similarity of this method to that discussed in the chapter on equilibrium.

Following these steps will get you majority credit on an AP free-response problem even if you do not ultimately get the correct answer. In fact, even if you only get through the first one or two steps, it is likely that you will still get some credit.

Only *Net* Force Equals *ma*

THIS IS REALLY IMPORTANT. Only F_{net} can be set equal to *ma*. You cannot set any old force equal to *ma*. For example, let's say that you have a block of mass *m* sitting on a table. The force of gravity, *mg*, acts down on the block. But that does not mean that you can say, "$F = mg$, so the acceleration of the block is *g*, or about 10 m/s²." Because we know that the block isn't falling! Instead, we know that the table exerts a normal force on the block that is equal in magnitude but opposite in direction to the force exerted by gravity. So the NET force acting on the block is 0. You can say "$F_{net} = 0$, so the block is not accelerating."

A Simple Example

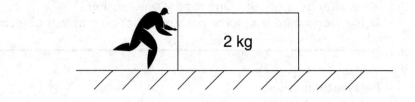

A block of mass *m* = 2 kg is pushed along a frictionless surface. The force pushing the block has a magnitude of 5 N and is directed at $\theta = 30°$ below the horizontal. What is the block's acceleration?

We follow our four-step process. First, draw a proper free-body diagram.

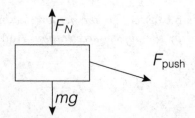

Second, we break the F_{push} vector into components that line up with the horizontal and vertical axes.

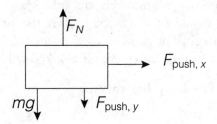

We can now move on to Step 3, writing equations for the net force in each direction:

$$F_{net,x} = (\text{right forces} - \text{left forces}) = ma_x$$

$$F_{net,x} = (F_{push,x} - 0) = ma_x$$

$$F_{net,y} = (\text{up forces} - \text{down forces}) = ma_y$$

$$F_{net,y} = (F_N - mg - F_{push,y}) = ma_y$$

Now we can plug our known values into these equations and solve for the acceleration of the block. First, we solve for the right–left direction:

$$F_{net,x} = (5\,\text{N})(\cos 30°) - 0 = (2\,\text{kg})a_x$$

$$a_x = 2.2\,\text{m/s}^2$$

Next, we solve for the up–down direction. Notice that the block is in equilibrium in this direction—it is neither flying off the table nor being pushed through it—so we know that the net force in this direction must equal 0.

$$F_{net,y} = 0 = (2\,\text{kg})a_y$$

$$a_y = 0$$

So the acceleration of the block is simply 2.2 m/s² to the right.

F_{net} on Inclines

A block of mass m is placed on a plane inclined at an angle θ. The coefficient of friction between the block and the plane is μ. What is the acceleration of the block down the plane?

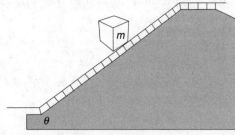

This is a really boring problem. But it's also a really common problem, so it's worth looking at.[1]

Note that no numbers are given, just variables. That's okay. It just means that our answer should be in variables. Only the given variables—in this case m, θ, and μ—and constants such as g can be used in the solution. And you shouldn't plug in any numbers (such as 10 m/s² for g), even if you know them.

Step 1: Free-body diagram.

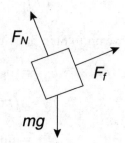

Step 2: Break vectors into components.

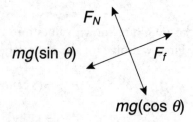

If you don't know where we got those vector components, refer back to Chapter 10.

Step 3: Write equations for the net force in each direction.
Note that the block is in equilibrium in the direction perpendicular to the plane, so the equation for $F_{\text{net, perpendicular}}$ (but not the equation for $F_{\text{net, down the plane}}$) can be set equal to 0.

$$F_{\text{net, perpendicular}} = mg(\cos\theta) - F_N = 0$$
$$mg(\cos\theta) = F_N$$
$$F_{\text{net, down the plane}} = mg(\sin\theta) - F_f = ma_{\text{down the plane}}$$

Step 4: Solve.
We can rewrite F_f, because

$$F_f = \mu F_N = \mu mg(\cos\theta)$$

Plugging this expression for F_f into the "$F_{\text{net, down the plane}}$" equation, we have

$$ma_{\text{down the plane}} = mg(\sin\theta - \mu\cos\theta)$$

Answer: $a_{\text{down the plane}} = g(\sin\theta - \mu\cos\theta)$

---

[1]If you want to make this problem more interesting, just replace the word "block" with the phrase "maniacal tobogganist" and the word "plane" with the phrase "highway on-ramp."

It always pays to check the reasonability of the answer. First, the answer doesn't include any variables that weren't given. Next, the units work out: g has acceleration units; neither the sine or cosine of an angle nor the coefficient of friction has any units.

Second, compare the answer to something familiar. Note that if the plane were vertical, $\theta = 90°$, so the acceleration would be g—yes, the block would then be in free fall! Also, note that friction tends to make the acceleration smaller, as you might expect.

For this particular incline, what coefficient of friction would cause the block to slide with constant speed?

Constant speed means $a = 0$. The solution for F_N in the perpendicular direction is the same as before: $F_N = mg(\cos\theta)$. But in the down-the-plane direction, no acceleration means that $F_f = mg(\sin\theta)$. Because $\mu = F_f / F_N$,

$$\mu = \frac{mg\sin\theta}{mg\cos\theta}$$

Canceling terms and remembering that sin/cos = tan, you find that $\mu = \tan\theta$ when acceleration is zero.

You might note that neither this answer nor the previous one includes the mass of the block, so on the same plane, both heavy and light masses move the same way!

F_{net} for a Pulley

Before we present our next practice problem, a few words about tension and pulleys are in order. Tension in a rope is the same *everywhere* in the rope, even if the rope changes direction (such as when it goes around a pulley) or if the tension acts in different directions on different objects. ONE ROPE = ONE TENSION. If there are multiple ropes in a problem, each rope will have its own tension. TWO ROPES = TWO TENSIONS.[2]

When masses are attached to a pulley, the pulley can only rotate one of two ways. Call one way positive, the other, negative.

A block of mass M and a block of mass m are connected by a thin string that passes over a light frictionless pulley. Find the acceleration of the system.

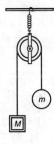

We arbitrarily call counterclockwise rotation of the pulley "positive."

---

[2]Except for the Physics C corollary, when the pulley is massive, but massive pulleys will not show up on the Physics B exam.

Step 1: Free-body diagrams.

The tension T is the same for each block—ONE ROPE = ONE TENSION. Also, note that because the blocks are connected, they will have the same acceleration, which we call *a*.

Step 2: Components.
The vectors already line up with one another. On to step 3.

Step 3: Equations.

$$\text{Block } M:\ F_{net} = Mg - T = Ma$$
$$\text{Block } m:\ F_{net} = T - mg = ma$$

Notice how we have been careful to adhere to our convention of which forces act in the positive and negative directions.

Step 4: Solve.
Let's solve for *T* using the first equation:

$$T = Mg - Ma$$

Plugging this value for *T* into the second equation, we have

$$(Mg - Ma) - mg = ma$$

Our answer is

$$a = g\left(\frac{M-m}{M+m}\right)$$

A 2-kg block and a 5-kg block are connected as shown on a frictionless surface. Find the tension in the rope connecting the two blocks. Ignore any friction effects.

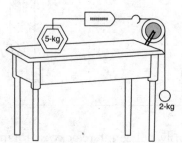

Why don't you work this one out for yourself? We have included our solution on the following page.

Solution to Example Problem

Step 1: Free-body diagrams.

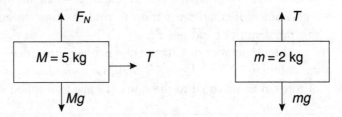

Step 2: Components.
Again, our vectors line up nicely, so on to Step 3.

Step 3: Equations.
Before we write any equations, we must be careful about signs: we shall call counterclockwise rotation of the pulley "positive."

For the more massive block, we know that, because it is not flying off the table or tunneling into it, it is in equilibrium in the up–down direction. But it is not in equilibrium in the right–left direction.

$$F_{net,y} = (F_N - Mg) = 0$$
$$F_{net,x} = (0 - T) = Ma$$

For the less massive block, we only have one direction to concern ourselves with: the up–down direction.

$$F_{net} = T - mg = ma$$

We can solve for T from the "$F_{net,x}$" equation for the more massive block and plug that value into the "F_{net}" equation for the less massive block, giving us

$$(-Ma) - mg = ma.$$

We rearrange some terms to get

$$a = \frac{-mg}{m + M}.$$

Now we plug in the known values for M and m to find that

$$a = -\frac{2}{7} g.$$

To finish the problem, we plug this value for a into the "$F_{net,x}$" equation for the more massive block.

$$-T = Ma$$
$$-T = (5)\left(-\frac{2}{7} g\right)$$
$$T = 14 \text{ N}$$

More Thoughts on $F_{net} = ma$

The four example problems in this chapter were all solved using only $F_{net} = ma$. Problems you might face in the real world—that is, on the AP test—will not always be so straightforward. Here's an example: imagine that this last example problem asked you to find the speed of the blocks after 2 seconds had elapsed, assuming that the blocks were released from rest. That's a kinematics problem, but to solve it, you have to know the acceleration of the blocks. You would first have to use $F_{net} = ma$ to find the acceleration, and then you could use a kinematics equation to find the final speed. We suggest that you try to solve this problem: it's good practice.

Also, remember in Chapter 9 when we introduced the unit of force, the newton, and we said that 1 N = 1 kg·m/s²? Well, now you know why that conversion works: the units of force must be equal to the units of mass multiplied by the units of acceleration.

Exam tip from an AP Physics veteran:
Newton's second law works for *all* kinds of forces, not just tensions, friction, and such. Often what looks like a complicated problem with electricity or magnetism is really just an $F_{net} = ma$ problem, but the forces might be electric or magnetic in nature.

—Jonas, high school senior

Newton's Third Law

We're sure you've been able to quote the third law since birth, or at least since 5th grade: "Forces come in equal and opposite action-reaction pairs," also known as "For every action there is an equal and opposite reaction." If I push down on the Earth, the Earth pushes up on me; a football player who makes a tackle experiences the same force that he dishes out.

What's so hard about that? Well, ask yourself one of the most important conceptual questions in first-year physics: "If all forces cause reaction forces, then how can anything ever accelerate?" Pull a little lab cart horizontally across the table . . . you pull on the cart, the cart pulls on you, so don't these forces cancel out, prohibiting acceleration?

Well, obviously, things can move. The trick is, Newton's third law force pairs *must act on different objects*, and so can never cancel each other.

When writing $F_{net} = ma$, only consider the forces acting on the object in question. Do not include forces exerted *by* the object.

Consider the lab cart. The only horizontal force that it experiences is the force of your pull. So, it accelerates toward you. Now, you experience a force from the cart, but you also experience a whole bunch of other forces that keep you in equilibrium; thus, you don't go flying into the cart.

› Practice Problems

Multiple Choice:

1. A 2.0-kg cart is given a shove up a long, smooth 30° incline. If the cart is traveling 8.0 m/s after the shove, how much time elapses until the cart returns to its initial position?

(A) 1.6 s
(B) 3.2 s
(C) 4.0 s
(D) 6.0 s
(E) 8.0 s

2. A car slides up a frictionless inclined plane. How does the normal force of the incline on the car compare with the weight of the car?

(A) The normal force must be equal to the car's weight.
(B) The normal force must be less than the car's weight.
(C) The normal force must be greater than the car's weight.
(D) The normal force must be zero.
(E) The normal force could have any value relative to the car's weight.

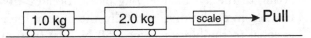

3. In the diagram above, a 1.0-kg cart and a 2.0-kg cart are connected by a rope. The spring scale reads 10 N. What is the tension in the rope connecting the two carts? Neglect any friction.

(A) 30 N
(B) 10 N
(C) 6.7 N
(D) 5.0 N
(E) 3.3 N

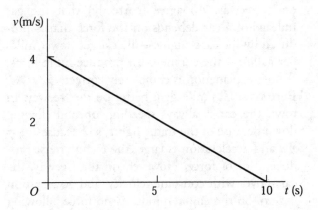

4. The velocity–time graph above represents the motion of a 5-kg box. The only force applied to this box is a person pushing. Assuming that the box is moving to the right, what is the magnitude and direction of the force applied by the person pushing?

(A) 2.0 N, right
(B) 2.0 N, left
(C) 0.4 N, right
(D) 0.4 N, left
(E) 12.5 N, left

Free Response:

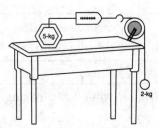

5. A 2-kg block and a 5-kg block are connected as shown above. The coefficient of friction between the 5-kg block and the flat surface is $\mu = 0.2$.

(A) Calculate the magnitude of the acceleration of the 5-kg block.
(B) Calculate the tension in the rope connecting the two blocks.

6. Bert, Ernie, and Oscar are discussing the gas mileage of cars. Specifically, they are wondering whether a car gets better mileage on a city street or on a freeway. All agree (correctly) that the gas mileage of a car depends on the force that is produced by the car's engine—the car gets fewer miles per gallon if the engine must produce more force. Whose explanation is completely correct?

Bert says: Gas mileage is better on the freeway. In town the car is always speeding up and slowing down because of the traffic lights, so because $F_{net} = ma$ and acceleration is large, the engine must produce a lot of force. However, on the freeway, the car moves with constant velocity, and acceleration is zero. So the engine produces no force, allowing for better gas mileage.

Ernie says: Gas mileage is better in town. In town, the speed of the car is slower than the speed on the freeway. Acceleration is velocity divided by time, so the acceleration in town is smaller. Because $F_{net} = ma$, then, the force of the engine is smaller in town giving better gas mileage.

Oscar says: Gas mileage is better on the freeway. The force of the engine only has to be enough to equal the force of air resistance—the engine doesn't have to accelerate the car because the car maintains a constant speed. Whereas in town, the force of the engine must often be greater than the force of friction and air resistance in order to let the car speed up.

› Solutions to Practice Problems

1. **B**—"Smooth" usually means, "ignore friction." So the only force acting along the plane is a component of gravity, $mg(\sin 30°)$. The F_{net} equation becomes $mg(\sin 30°) - 0 = ma$. The mass cancels, leaving the acceleration as 5 m/s². What's left is a kinematics problem. Set up a chart, calling the direction down the plane as positive:

v_0	−8.0/s
v_f	?
$x - x_0$	0 (cart comes back to starting point)
a	+5.0 m/s²
t	??? what we're looking for

Use ** $(\Delta x = v_0 t + \frac{1}{2}at^2)$ to find that the time is 3.2 s.

2. **B**—The normal force exerted on an object on an inclined plane equals $mg(\cos \theta)$, where θ is the angle of the incline. If θ is greater than 0, then $\cos \theta$ is less than 1, so the normal force is less than the object's weight.

3. **E**—Consider the forces acting on each block separately. On the 1.0-kg block, only the tension acts, so $T = (1.0 \text{ kg})a$. On the 2.0-kg block, the tension acts left, but the 10 N force acts right, so $10 \text{ N} - T = (2.0 \text{ kg})a$. Add these equations together (noting that the tension in the rope is the same in both equations), getting $10 \text{ N} = (3.0 \text{ kg})a$; acceleration is 3.3 m/s². To finish, $T = (1.0 \text{ kg})a$, so tension is 3.3 N.

4. **B**—The acceleration is given by the slope of the v–t graph, which has magnitude 0.4 m/s². $F_{net} = ma$, so 5 kg × 0.4 m/s² = 2.0 N. This force is to the left because acceleration is negative (the slope is negative), and negative was defined as left.

5. The setup is the same as in the chapter's example problem, except this time there is a force of friction acting to the left on the 5-kg block. Because this block is in equilibrium vertically, the normal force is equal to the block's weight, 50 N. The friction force is μF_N, or 10 N.

   Calling the down-and-right direction positive, we can write two equations, one for each block:

$$(2 \text{ kg})g - T = (2 \text{ kg})a$$
$$T - F_f = (5 \text{ kg})a$$

   (A) To solve for acceleration, just add the two equations together. The tensions cancel. We find the acceleration to be 1.4 m/s².

   (B) Plug back into either equation to find the final answer, that the tension is 17 N. This is more than the 14 N we found for the frictionless situation, and so makes sense. We expect that it will take more force in the rope to overcome friction on the table.

6. Although Bert is right that acceleration is zero on the freeway, this means that the NET force is zero; the engine still must produce a force to counteract air resistance. This is what Oscar says, so his answer is correct. Ernie's answer is way off—acceleration is not velocity/time, acceleration is a CHANGE in velocity over time.

› Rapid Review

- The net force on an object equals the mass of the object multiplied by the object's acceleration.

- To solve a problem using $F_{net} = ma$, start by drawing a good free-body diagram. Resolve forces into vector components. For each axis, the vector sum of forces along that axis equals ma_i, where a_i is the acceleration of the object along that axis.

- When an object is on an inclined plane, resolve its weight into vector components that point parallel and perpendicular to the plane.

- For problems that involve a pulley, remember that if there's one rope, there's one tension.

CHAPTER 13

Momentum

IN THIS CHAPTER

Summary: The impulse–momentum relationship can explain how force acts in a collision. Momentum is conserved in all collisions, allowing a prediction of objects' speeds before and after a collision.

Key Ideas

- ✪ Impulse can be expressed both as force times a time interval, *and* as a change in momentum.
- ✪ The total momentum of a set of objects before a collision is equal to the total momentum of a set of objects after a collision.
- ✪ Momentum is a vector, so leftward momentum can "cancel out" rightward momentum.

Relevant Equations

The definition of momentum:

$$p = mv$$

The impulse–momentum theorem:

$$\Delta p = F\Delta t$$

If an object is moving, it has momentum. The formal definition of momentum is that it's equal to an object's mass multiplied by that object's velocity. However, a more intuitive way to think about momentum is that it corresponds to the amount of "oomph" an object has in a collision. Regardless of how you think about momentum, the key is this: the momentum of a system upon which no net external force acts is always conserved.

Momentum and Impulse

$$\boxed{\textbf{Momentum: } mv}$$

The units of momentum are kg·m/s which is the same as N·s. Momentum is a vector quantity, and it is often abbreviated with a p.

$$\boxed{\textbf{Impulse: } \Delta p = F\Delta t}$$

Impulse (designated as I) is an object's change in momentum. It is also equal to the force acting on an object multiplied by the time interval over which that force was applied. The above equation is often referred to as the "impulse–momentum theorem."

The $F\Delta t$ definition of impulse explains why airbags are used in cars and why hitting someone with a pillow is less dangerous than hitting him or her with a cement block. The key is the Δt term. An example will help illustrate this point.

A man jumps off the roof of a building, 3.0 m above the ground. His mass is 70 kg. He estimates (while in free fall) that if he lands stiff-legged, it will take him 3 ms (milliseconds) to come to rest. However, if he bends his knees upon impact, it will take him 100 ms to come to rest. Which way will he choose to land, and why?

This is a multistep problem. We start by calculating the man's velocity the instant before he hits the ground. That's a kinematics problem, so we start by choosing a positive direction—we'll choose "down" to be positive—and by writing out our table of variables.

v_0	0
v_f	?
Δx	3.0 m
a	10 m/s^2
t	?

We have three variables with known values, so we can solve for the other two. We don't care about time, t, so we will just solve for v_f.

$$v_f^2 = v_0^2 + 2a(x - x_0)$$
$$v_f = 7.7 \text{ m/s}$$

Now we can solve for the man's momentum the instant before he hits the ground.

$$p = mv = (70)(7.7) = 540 \text{ kg·m/s}$$

Once he hits the ground, the man quickly comes to rest. That is, his momentum changes from 540 kg·m/s to 0.

$$I = \Delta p = P_f - p_0$$
$$I = -540 \text{ N·s} = F\Delta t$$

If the man does not bend his knees, then

$$-540 = F(0.003 \text{ s})$$
$$F = -180,000 \text{ N}$$

The negative sign in our answer just means that the force exerted on the man is directed in the negative direction: up.

Now, what if he had bent his knees?

$$I = -540 = F\Delta t$$
$$-540 = F(0.10 \text{ seconds})$$
$$F = -5400 \text{ N}$$

If he bends his knees, he allows for his momentum to change more slowly, and as a result, the ground exerts a lot less force on him than had he landed stiff-legged. More to the point, hundreds of thousands of newtons applied to a person's legs will cause major damage—this is the equivalent of almost 20 tons sitting on his legs. So we would assume that the man would bend his knees upon landing, reducing the force on his legs by a factor of 30.

Conservation of Momentum

Momentum in an isolated system, where no net external forces act, is always conserved. A rough approximation of a closed system is a billiard table covered with hard tile instead of felt. When the billiard balls collide, they transfer momentum to one another, but the total momentum of all the balls remains constant.

The key to solving conservation of momentum problems is remembering that momentum is a vector.

A satellite floating through space collides with a small UFO. Before the collision, the satellite was traveling at 10 m/s to the right, and the UFO was traveling at 5 m/s to the left. If the satellite's mass is 70 kg, and the UFO's mass is 50 kg, and assuming that the satellite and the UFO bounce off each other upon impact, what is the satellite's final velocity if the UFO has a final velocity of 3 m/s to the right?

Let's begin by drawing a picture.

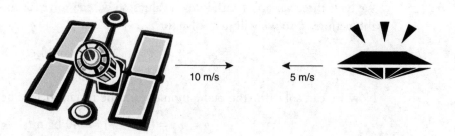

Momentum is conserved, so we write

$$p_{\text{satellite}} + p_{\text{UFO}} = p'_{\text{satellite}} + p'_{\text{UFO}}$$
$$(m_s)(v_s) + (m_{\text{UFO}})(v_{\text{UFO}}) = (m_s)(v_s)' + (m_{\text{UFO}})(v_{\text{UFO}})'.$$

The tick marks on the right side of the equation mean "after the collision." We know the momentum of each space traveler before the collision, and we know the UFO's final momentum. So we solve for the satellite's final velocity. (Note that we must define a positive direction; because the UFO is moving to the left, its velocity is plugged in as negative.)

$$(70 \text{ kg})(+10 \text{ m/s}) + (50 \text{ kg})(-5 \text{ m/s}) = (70 \text{ kg}) v_s' + (50 \text{ kg})(+3 \text{ m/s})$$
$$v_s' = 4.3 \text{ m/s to the right}$$

Now, what if the satellite and the UFO had stuck together upon colliding? We can solve for their final velocity easily:

$$p_{\text{satellite}} + p_{\text{UFO}} = p'_{\text{satellite \& UFO}}$$
$$(m_s)(v_s) + (m_{\text{UFO}})(v_{\text{UFO}}) = (m_{s \text{ \& UFO}})(v_{s \text{ \& UFO}})'$$
$$(70 \text{ kg})(+10 \text{ m/s}) + (50 \text{ kg})(-5 \text{ m/s}) = (70 \text{ kg} + 50 \text{ kg})(v_{s \text{ \& UFO}})'$$
$$v_{s \text{ \& UFO}}' = 3.8 \text{ m/s to the right}$$

Elastic and Inelastic Collisions

This brings us to a couple of definitions.

> **Elastic Collision:** A collision in which kinetic energy is conserved

If you're unfamiliar with the concept of kinetic energy (KE), take a few minutes to skim Chapter 14 right now.

When the satellite and the UFO bounced off each other, they experienced a perfectly elastic collision. If kinetic energy is lost to heat or anything else during the collision, it is called an inelastic collision.

> **Inelastic Collision:** A collision in which KE is not conserved

The extreme case of an inelastic collision is called a perfectly inelastic collision.

> **Perfectly Inelastic Collision:** The colliding objects stick together after impact

The second collision between the satellite and the UFO was a perfectly inelastic collision. BUT, MOMENTUM IS STILL CONSERVED, EVEN IN A PERFECTLY INELASTIC COLLISION!

Two-Dimensional Collisions

The key to solving a two-dimensional collision problem is to remember that momentum is a vector, and as a vector it can be broken into x and y components. Momentum in the x-direction is always conserved, and momentum in the y-direction is always conserved.

Maggie has decided to go ice-skating. While cruising along, she trips on a crack in the ice and goes sliding. She slides along the ice at a velocity of 2.5 m/s. In her path is a penguin. Unable to avoid the flightless bird, she collides with it. The penguin is initially at rest and has a mass of 20 kg, and Maggie's mass is 50 kg. Upon hitting the penguin, Maggie is deflected 30° from her initial path, and the penguin is deflected 60° from Maggie's initial path. What is Maggie's velocity, and what is the penguin's velocity, after the collision?

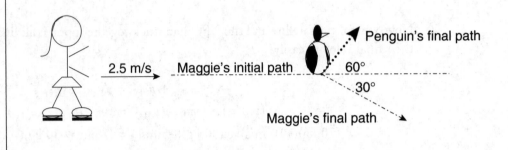

We want to analyze the *x*-component of momentum and the *y*-component of momentum separately. Let's begin by defining "right" and "up" to be the positive directions. Now we can look at the *x*-component.

$$(m_{\text{Maggie}})(v_{\text{Maggie},x}) + (m_{\text{penguin}})(v_{\text{penguin},x}) = (m_M)(v_{M,x})' + (m_p)(v_{p,x})'$$
$$(50 \text{ kg})(+2.5 \text{ m/s}) + (20 \text{ kg})(0) = (50 \text{ kg})(+v'_M \cos 30°) + (20 \text{ kg})(+v'_p \cos 60°)$$
$$125 \text{ kg·m/s} = (50 \text{ kg})(+v'_M)(\cos 30°) + (20 \text{ kg})(+v'_p)(\cos 60°)$$

We can't do much more with the *x*-component of momentum, so now let's look at the *y*-component.

$$(m_{\text{Maggie}})(v_{\text{Maggie}, y}) + (m_{\text{penguin}})(v_{\text{penguin}, y}) = (m_M)(v_{M,y})' + (m_p)(v_{p,y})'$$
$$(50 \text{ kg})(0) + (20 \text{ kg})(0) = (50 \text{ kg})(-v'_M \sin 30°) + (20 \text{ kg})(+v'_p \sin 60°)$$

(Note the negative sign on Maggie's *y*-velocity!)

$$0 = (50 \text{ kg})(-v'_M \sin 30°) + (20 \text{ kg})(+v'_p \sin 60°)$$

Okay. Now we have two equations and two unknowns. It'll take some algebra to solve this one, but none of it is too hard. We will assume that you can do the math on your own, but we will gladly provide you with the answer:

$$v'_M = 2.2 \text{ m/s}$$
$$v'_p = 3.1 \text{ m/s}$$

The algebra is not particularly important here. Get the conceptual physics down—in a two-dimensional collision, you must treat each direction separately. If you do so, you will receive virtually full credit on an AP problem. If you combine vertical and horizontal momentum into a single conservation equation, you will probably not receive any credit at all.

❯ Practice Problems

Multiple Choice:

First two questions: A ball of mass M is caught by someone wearing a baseball glove. The ball is in contact with the glove for a time t; the initial velocity of the ball (just before the catcher touches it) is v_0.

1. If the time of the ball's collision with the glove is doubled, what happens to the force necessary to catch the ball?

   (A) It doesn't change.
   (B) It is cut in half.
   (C) It is cut to one fourth of the original force.
   (D) It quadruples.
   (E) It doubles.

2. If the time of collision remains t, but the initial velocity is doubled, what happens to the force necessary to catch the ball?

   (A) It doesn't change.
   (B) It is cut in half.
   (C) It is cut to one fourth of the original force.
   (D) It quadruples.
   (E) It doubles.

before collision

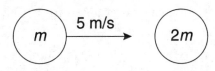

3. Two balls, of mass m and $2m$, collide and stick together. The combined balls are at rest after the collision. If the ball of mass m was moving 5 m/s to the right before the collision, what was the velocity of the ball of mass $2m$ before the collision?

   (A) 2.5 m/s to the right
   (B) 2.5 m/s to the left
   (C) 10 m/s to the right
   (D) 10 m/s to the left
   (E) 1.7 m/s to the left

4. Two identical balls have initial velocities $v_1 = 4$ m/s to the right and $v_2 = 3$ m/s to the left, respectively. The balls collide head on and stick together. What is the velocity of the combined balls after the collision?

   (A) ½ m/s to the right
   (B) ⅔ m/s to the right
   (C) ½ m/s to the right
   (D) ⅖ m/s to the right
   (E) 1 m/s to the right

Free Response:

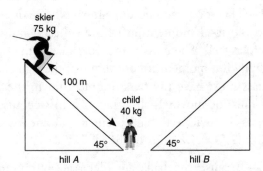

5. A 75-kg skier skis down a hill. The skier collides with a 40-kg child who is at rest on the flat surface near the base of the hill, 100 m from the skier's starting point, as shown above. The skier and the child become entangled. Assume all surfaces are frictionless.

   (a) How fast will the skier be moving when he reaches the bottom of the hill? Assume the skier is at rest when he begins his descent.
   (b) What will be the speed of the skier and child just after they collide?
   (c) If the collision occurs in half a second, how much force will be experienced by each person?

› Solutions to Practice Problems

1. **B**—Impulse is force times the time interval of collision, and is also equal to an object's change in momentum. Solving for force, $F = \Delta p/\Delta t$. Because the ball still has the same mass, and still changes from speed v_0 to speed zero, the ball's momentum change is the same, regardless of the collision time. The collision time, in the denominator, doubled; so the entire expression for force was cut in half.

2. **E**—Still use $F = \Delta p/\Delta t$, but this time it is the numerator that changes. The ball still is brought to rest by the glove, and the mass of the ball is still the same; but the doubled velocity upon reaching the glove doubles the momentum change. Thus, the force doubles.

3. **B**—The total momentum after collision is zero. So the total momentum before collision must be zero as well. The mass m moved 5 m/s to the right, giving it a momentum of $5m$ units; the right-hand mass must have the same momentum to the left. It must be moving half as fast, 2.5 m/s because its mass is twice as big; then its momentum is $(2m)(2.5) = 5m$ units to the left.

4. **C**—Because the balls are identical, just pretend they each have mass 1 kg. Then the momentum conservation tells us that

   $$(1 \text{ kg})(+4 \text{ m/s}) + (1 \text{ kg})(-3 \text{ m/s}) = (2 \text{ kg})(v').$$

   The combined mass, on the right of the equation above, is 2 kg; v' represents the speed of the combined mass. Note the negative sign indicating the direction of the second ball's velocity. Solving, $v' = +0.5$ m/s, or 0.5 m/s to the right.

5. (a) This part is not a momentum problem, it's a Newton's second law and kinematics proplem. (Or it's an energy problem, if you've studied energy.) Break up forces on the skier into parallel and perpendicular axes—the net force down the plane is $mg(\sin 45°)$. So by Newton's second law, the acceleration down the plane is g $(\sin 45°) = 7.1$ m/s^2. Using kinematics with intitial velocity zero and distance 100 m, the skier is going 38 m/s (!).

   (b) Now use momentum conservation. The total momentum before collision is (75 kg)(38 m/s) = 2850 kg·m/s. This must equal the total momentum after collision. The people stick together, with combined mass 115 kg. So after collision, the velocity is 2850 kg·m/s divided by 115 kg, or about 25 m/s.

   (c) Change in momentum is force multiplied by time interval . . . the child goes from zero momentum to (40 kg)(25 m/s) = 1000 kg·m/s of momentum. Divide this change in momentum by 0.5 seconds, and you get 2000 N, or a bit less than a quarter ton of force. Ouch!

› Rapid Review

- Momentum equals an object's mass multiplied by its velocity. However, you can also think of momentum as the amount of "oomph" a mass has in a collision.

- Impulse equals the change in an object's momentum. It also equals the force exerted on an object multiplied by the time it took to apply that force.

- Momentum is always conserved. When solving conservation of momentum problems, remember that momentum is a vector quantity.

- In an elastic collision, kinetic energy is conserved. When two objects collide and bounce off each other, without losing any energy (to heat, sound, etc.), they have engaged in an elastic collision. In an inelastic collision, kinetic energy is not conserved. The extreme case is a perfectly inelastic collision. When two objects collide and stick together, they have engaged in a perfectly inelastic collision.

CHAPTER 14

Energy Conservation

IN THIS CHAPTER

Summary: While kinematics can be used to predict the speeds of objects with constant acceleration, energy conservation is a more powerful tool that can predict how objects will move even with a changing acceleration.

Key Ideas

✪ Work is related to kinetic energy through the work–energy theorem.
✪ There are many types of potential energy. Two (due to gravity and due to a spring) are discussed in this chapter.
✪ To use conservation of energy, add potential + kinetic energy at two positions in an object's motion. This sum must be the same everywhere.
✪ Physics B students also have to know about electrical potential energy, as covered in Chapter 20; but the problem-solving approach for energy conservation is always the same.

Relevant Equations

The definition of work:

$$W = F \cdot d_{\parallel}$$

The work–energy theorem:

$$W_{net} = \Delta KE$$

The force of a spring:

$$F = kx$$

Two different types of potential energy:

$$\text{Gravitational PE} = mgh$$
$$\text{Spring PE} = \tfrac{1}{2}kx^2$$

Power:

$$\text{Power} = \frac{\text{work}}{\text{time}} = Fv$$

As with momentum, the energy of an isolated system is always conserved. It might change form—potential energy can be converted to kinetic energy, or kinetic energy can be converted to heat—but it'll never simply disappear.

Conservation of energy is one of the most important, fundamental concepts in all of physics . . . translation: it's going to show up all over the AP exam. So read this chapter carefully.

Kinetic Energy and the Work–Energy Theorem

We'll start with some definitions.

> **Energy:** The ability to do work

> **Work:** $F \cdot d_{\parallel}$

What this second definition means is that work equals the product of the distance an object travels and the component of the force acting on that object directed parallel to the object's direction of motion. That sounds more complicated than it really is: an example will help.

A box is pulled along the floor, as shown in Figure 14.1. It is pulled a distance of 10 m, and the force pulling it has a magnitude of 5 N and is directed 30° above the horizontal. So, the force component that is PARALLEL to the 10 m displacement is (5 N)(cos 30°).

$$W = (5 \cos 30° \text{ N})(10 \text{ m})$$
$$W = 43 \text{ N} \cdot \text{m}$$

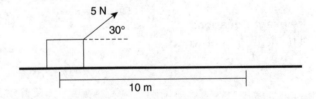

Figure 14.1   A box pulled along the floor.

One newton·meter is called a joule, abbreviated as 1 J.

- Work is a scalar. So is energy.
- The units of work and of energy are joules.
- Work can be negative . . . this just means that the force is applied in the direction opposite displacement.

> **Kinetic Energy:** Energy of motion, abbreviated KE

$$KE = \tfrac{1}{2}mv^2$$

This means that the kinetic energy of an object equals one half the object's mass times its speed squared.

> **Work–Energy Theorem:** $W_{net} = \Delta KE$

The net work done on an object is equal to that object's change in kinetic energy. Here's an application:

> A train car with a mass of 200 kg is traveling at 20 m/s. How much force must the brakes exert in order to stop the train car in a distance of 10 m?
>
>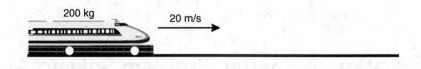

Here, because the only horizontal force is the force of the brakes, the work done by this force is W_{net}.

$$W_{net} = \Delta KE = KE_f - KE_0$$

$$W = \tfrac{1}{2}mv_f^2 - \tfrac{1}{2}mv_0^2$$

$$W = (0) - \tfrac{1}{2}(200 \text{ kg})(20 \text{ m/s})^2$$

$$W = -40{,}000 \text{ J}$$

Let's pause for a minute to think about what this value means. We've just calculated the change in kinetic energy of the train car, which is equal to the net work done on the train car. The negative sign simply means that the net force was opposite the train's displacement. To find the force:

$$-40{,}000 \text{ N} = \text{Force} \cdot \text{distance}$$

$$-40{,}000 \text{ N} = F(10 \text{ m})$$

$$F = -4000 \text{ N, which means 4000 N opposite the}$$
$$\text{direction of displacement}$$

Potential Energy

> **Potential Energy:** Energy of position, abbreviated PE

Potential energy comes in many forms: there's gravitational potential energy, spring potential energy, electrical potential energy, and so on. For starters, we'll concern ourselves with gravitational potential energy.

Gravitational PE is described by the following equation:

$$PE = mgh$$

In this equation, m is the mass of an object, g is the gravitational field of 10 N/kg on Earth, and h is the height of an object above a certain point (called "the zero of potential").[1] That point can be wherever you want it to be, depending on the problem. For example, let's say a pencil is sitting on a table. If you define the zero of potential to be the table, then the pencil has no gravitational PE. If you define the floor to be the zero of potential, then the pencil has PE equal to mgh, where h is the height of the pencil above the floor. Your choice of the *zero* of potential in a problem should be made by determining how the problem can most easily be solved.

REMINDER: h in the potential energy equation stands for *vertical* height above the zero of potential.

Conservation of Energy: Problem-Solving Approach

Solving energy-conservation problems is relatively simple, as long as you approach them methodically. The general approach is this: write out all the terms for the initial energy of the system, and set the sum of those terms equal to the sum of all the terms for the final energy of the system. Let's practice.

A block of mass m is placed on a frictionless plane inclined at a 30° angle above the horizontal. It is released from rest and allowed to slide 5 m down the plane. What is its final velocity?

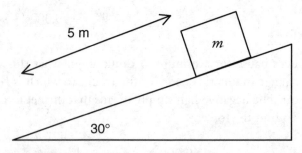

If we were to approach this problem using kinematics equations (which we could), it would take about a page of work to solve. Instead, observe how quickly it can be solved using conservation of energy.

---

[1] Note that 10 N/kg is exactly the same as 10 m/s².

$$KE_i + PE_i = KE_f + PE_f \begin{pmatrix} i = \text{initial, at the top} \\ f = \text{final, at the bottom} \end{pmatrix}$$

$$\tfrac{1}{2}mv_i^2 + mgh_i = \tfrac{1}{2}mv_f^2 + mgh_f$$

We will define our zero of potential to be the height of the box after it has slid the 5 m down the plane. By defining it this way, the PE term on the right side of the equation will cancel out. Furthermore, because the box starts from rest, its initial KE also equals zero.

$$0 + mgh_i = \tfrac{1}{2}mv_f^2 + 0$$

The initial height can be found using trigonometry: $h_i = (5m)(\sin 30°) = 2.5$ m.

$$mg(2.5\,\text{m}) = \tfrac{1}{2}mv_f^2$$
$$50(\text{m/s})^2 = v_f^2$$
$$v_f = 7\,\text{m/s}$$

In general, the principle of energy conservation can be stated mathematically like this:

$$\boxed{KE_i + PE_i + W = KE_f + PE_f}$$

The term W in this equation stands for work done on an object. For example, if there had been friction between the box and the plane in the previous example, the work done by friction would be the W term. When it comes to the AP exam, *you will include this W term only when there is friction (or some other exteral force) involved*. When friction is involved, $W = F_f d$, where F_f is the force of friction on the object, and d is the distance the object travels.

Let's say that there was friction between the box and the inclined plane.

A box of mass m is placed on a plane inclined at a 30° angle above the horizontal. The coefficient of friction between the box and the plane is 0.20. The box is released from rest and allowed to slide 5.0 m down the plane. What is its final velocity?

We start by writing the general equation for energy conservation:

$$KE_i + PE_i + W = KE_f + PE_f$$

W equals $F_f d$, where F_f is the force of friction, and d is 5 m.[2]

$$W = -F_f d = -\mu F_N d$$
$$F_N = (mg\cos\theta), \text{where } \theta = 30°$$

The value for W is negative because friction acts opposite displacement. You may want to draw a free-body diagram to understand how we derived this value for F_N.
Now, plugging in values we have

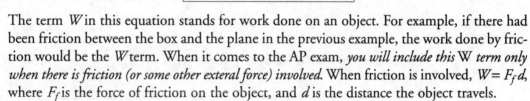

$$\tfrac{1}{2}mv_i^2 + mgh_i - 0.2(mg\cos 30°)(5\text{m}) = \tfrac{1}{2}mv_f^2 + mgh_f$$

$$0 + mg(2.5\,\text{m}) - 0.2(mg)(.87)(5\text{m}) = \tfrac{1}{2}mv_f^2$$

[2] Note this difference carefully. Although potential energy involves only a vertical height, work done by friction includes the *entire* distance the box travels.

We rearrange some terms and cancel out m from each side to get

$$v_f = 5.7 \text{ m/s}$$

This answer makes sense—friction on the plane *reduces* the box's speed at the bottom.

Springs

Gravitational potential energy isn't the only kind of PE around. Another frequently encountered form is spring potential energy.

The force exerted by a spring is directly proportional to the amount that the spring is compressed. That is,

$$\boxed{F_{\text{spring}} = kx}$$

In this equation, k is a constant (called the spring constant), and x is the distance that the spring has been compressed or extended from its equilibrium state.[3]

When a spring is either compressed or extended, it stores potential energy. The amount of energy stored is given as follows.

$$\boxed{PE_{\text{spring}} = \tfrac{1}{2}kx^2}$$

Similarly, the work done by a spring is given by $W_{\text{spring}} = \tfrac{1}{2}kx^2$. Here's an example problem.

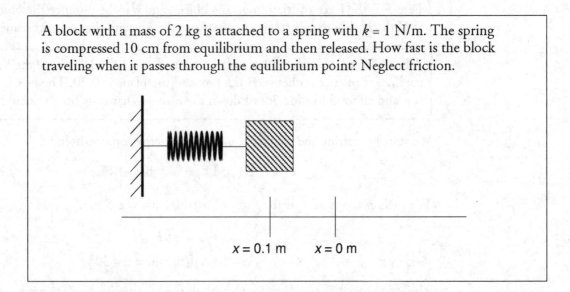

A block with a mass of 2 kg is attached to a spring with $k = 1$ N/m. The spring is compressed 10 cm from equilibrium and then released. How fast is the block traveling when it passes through the equilibrium point? Neglect friction.

$x = 0.1$ m $x = 0$ m

It's important to recognize that we CANNOT use kinematics to solve this problem! Because the force of a spring changes as it stretches, the block's acceleration is not constant. When acceleration isn't constant, try using energy conservation.

---

[3] On the equation sheet for the exam, you will see this equation written as $F = -kx$. The negative sign is simply a reminder that the force of a spring always acts opposite to displacement—an extended spring pulls back toward the equilibrium position, while a compressed spring pushes toward the equilibrium position. We call this type of force a restoring force, and we discuss it more in Chapter 16 about simple harmonic motion.

We begin by writing our statement for conservation of energy.

$$KE_i + PE_i + W = KE_f + PE_f$$

Now we fill in values for each term. PE here is just in the form of spring potential energy, and there's no friction, so we can ignore the W term. Be sure to plug in all values in meters!

$$0 + \tfrac{1}{2}(k)(0.1\,\text{m})^2 = \tfrac{1}{2}mv_f^2 + 0$$

$$v_f = \sqrt{\frac{k}{m}(0.1\,\text{m})^2}$$

Plugging in values for k and m, we have

$$v_f = 0.07 \text{ m/s, that is, 7 cm/s}$$

Power

Whether you walk up a mountain or whether a car drives you up the mountain, the same amount of work has to be done on you. (You weigh a certain number of newtons, and you have to be lifted up the same distance either way!) But clearly there's something different about walking up over the course of several hours and driving up over several minutes. That difference is power.

> **Power:** work/time

Power is, thus, measured in units of joules/second, also known as watts. A car engine puts out hundreds of horsepower, equivalent to maybe 100 kilowatts; whereas, you'd work hard just to put out a power of a few hundred watts.

› Practice Problems

Multiple Choice:

Questions 1 and 2

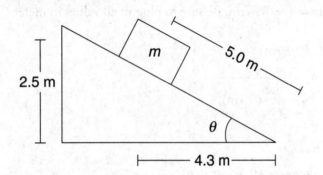

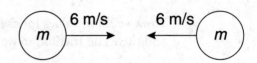

A block of weight $mg = 100$ N slides a distance of 5.0 m down a 30-degree incline, as shown above.

1. How much work is done on the block by gravity?

(A) 500 J
(B) 430 J
(C) 100 J
(D) 50 J
(E) 250 J

2. If the block experiences a constant friction force of 10 N, how much work is done by the friction force?

(A) −43 J
(B) −25 J
(C) −500 J
(D) −100 J
(E) −50 J

3. Two identical balls of mass $m = 1.0$ kg are moving towards each other, as shown above. What is the initial kinetic energy of the system consisting of the two balls?

(A) 0 joules
(B) 1 joules
(C) 12 joules
(D) 18 joules
(E) 36 joules

Free Response:

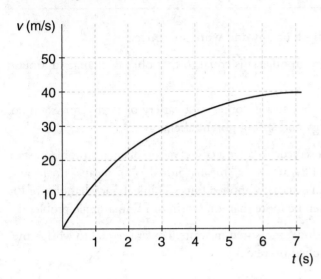

4. A 1500-kg car moves north according to the velocity–time graph shown.

   (a) Determine the change in the car's kinetic energy during the first 7 s.
   (b) To determine how far the car traveled in these 7 s, the three basic kinematics equations can not be used. Explain why not.
   (c) Use the velocity–time graph to estimate the distance the car traveled in 7 s.
   (d) What was the net work done on the car in these 7 s?
   (e) Determine the average power necessary for the car to perform this motion.

› Solutions to Practice Problems

1. **E**—The force of gravity is straight down and equal to 100 N. The displacement parallel to this force is the *vertical* displacement, 2.5 m. Work equals force times parallel displacement, 250 J.

2. **E**—The force of friction acts up the plane, and displacement is down the plane, so just multiply force times distance to get 50 J. The negative sign indicates that force is opposite displacement.

3. **E**—Kinetic energy is a scalar, so even though the balls move in opposite directions, the KEs cannot cancel. Instead, kinetic energy $\frac{1}{2}(1\text{ kg})(6\text{ m/s})^2$ attributable to different objects adds together algebraically, giving 36 J total.

4. (a) The car started from rest, or zero KE. The car ended up with $\frac{1}{2}(1500\text{ kg})(40\text{ m/s})^2 = 1.2 \times 10^6$ J of kinetic energy. So its change in KE is 1.2×10^6 J.

   (b) The acceleration is not constant. We know that because the velocity–time graph is not linear.

   (c) The distance traveled is found from the area under the graph. It is easiest to approximate by counting boxes, where one of the big boxes is 10 m. There are, give-or-take, 19 boxes underneath the curve, so the car went 190 m.

   (d) We cannot use work = force × distance here, because the net force is continually changing (because acceleration is changing). But $W_{net} = \Delta KE$ is always valid. In part (a) the car's change in KE was found to be 1.2×10^6 J; so the net work done on the car is also 1.2×10^6 J.

   (e) Power is work divided by time, or 1.2×10^6 J/7 s = 170 kW. This can be compared to the power of a car, 220 horsepower.

❯ Rapid Review

- Energy is the ability to do work. Both energy and work are scalars.

- The work done on an object (or by an object) is equal to that object's change in kinetic energy.

- Potential energy is energy of position, and it comes in a variety of forms; for example, there's gravitational potential energy and spring potential energy.

- The energy of a closed system is conserved. To solve a conservation of energy problem, start by writing $KE_i + PE_i + W = KE_f + PE_f$, where "$i$" means "initial," "$f$" means "final," and W is the work done by friction or an externally applied force. Think about what type of PE you're dealing with; there might even be more than one form of PE in a single problem!

- Power is the rate at which work is done, measured in watts. Power is equal to work/time, which is equivalent to force multiplied by velocity.

CHAPTER 15

Gravitation and Circular Motion

IN THIS CHAPTER

Summary: When an object moves in a circle, it is accelerating toward the center of the circle. Any two massive objects attract each other due to gravity.

Key Ideas

- ✪ Centripetal (center-seeking) acceleration is equal to $\frac{v^2}{r}$.
- ✪ Circular motion (and gravitation) problems still require a free-body diagram and the four-step problem-solving process.
- ✪ The gravitational force between two objects is bigger for bigger masses, and smaller for larger separations between the objects.

Relevant Equations

Centripetal acceleration:

$$a = \frac{v^2}{r}$$

Gravitational force between any two masses:

$$F = \frac{Gm_1 m_2}{r^2}$$

Gravitational potential energy a long way from a planet:

$$PE = \frac{Gm_1 m_2}{r}$$

It might seem odd that we're covering gravitation and circular motion in the same chapter. After all, one of these topics relates to the attractive force exerted between two massive objects, and the other one relates to, well, swinging a bucket over your head.

However, as you're probably aware, the two topics have plenty to do with each other. Planetary orbits, for instance, can be described only if you understand both gravitation and circular motion. And, sure enough, the AP exam frequently features questions about orbits.

So let's look at these two important topics, starting with circular motion.

Velocity and Acceleration in Circular Motion

Remember how we defined acceleration as an object's change in velocity in a given time? Well, velocity is a vector, and that means that an object's velocity can change either in magnitude or in direction (or both). In the past, we have talked about the magnitude of an object's velocity changing. Now, we discuss what happens when the direction of an object's velocity changes.

When an object maintains the same speed but turns in a circle, the magnitude of its acceleration is constant and directed **toward the center** of the circle. This means that the acceleration vector is perpendicular to the velocity vector at any given moment, as shown in Figure 15.1.

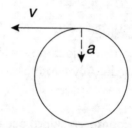

Figure 15.1   Velocity and acceleration of an object traveling in uniform circular motion.

The velocity vector is always directed tangent to the circle, and the acceleration vector is always directed toward the center of the circle. There's a way to prove that statement mathematically, but it's complicated, so you'll just have to trust us. (You can refer to your textbook for the complete derivation.)

Centripetal Acceleration

On to a few definitions.

> **Centripetal Acceleration:** The acceleration keeping an object in uniform circular motion, abbreviated a_c

We know that the net force acting on an object is related to the object's acceleration by $F_{net} = ma$. And we know that the acceleration of an object in circular motion points toward the center of the circle. So we can conclude that the centripetal force acting on an object also points toward the center of the circle.

The formula for centripetal acceleration is

$$a_c = \frac{v^2}{r}$$

In this equation, v is the object's velocity, and r is the radius of the circle in which the object is traveling.

> **Centrifugal Acceleration:** As far as you're concerned, nonsense. Acceleration in circular motion is always **toward**, not away from, the center.

Centripetal acceleration is real; centrifugal acceleration is nonsense, unless you're willing to read a multipage discussion of "non-inertial reference frames" and "fictitious forces." So for our purposes, there is no such thing as a centrifugal (center-fleeing) acceleration. *When an object moves in a circle, the acceleration through the net force on the object must act toward the center of the circle.*

The main thing to remember when tackling circular motion problems is that a *centripetal force is simply whatever force is directed toward the center of the circle in which the object is traveling.* So, first label the forces on your free-body diagram, and then find the net force directed toward the center of the circle. That net force is the centripetal force. But NEVER label a free-body diagram with "F_c."

> **Exam tip from an AP Physics veteran:**
> On a free-response question, do not label a force as "centripetal force," even if that force does act toward the center of a circle; you will not earn credit. Rather, label with the actual source of the force; i.e., tension, friction, weight, electric force, etc.
>
> *—Mike, high school junior*

Mass on a String

> A block of mass $M = 2$ kg is swung on a rope in a vertical circle of radius r at constant speed v. When the block is at the top of the circle, the tension in the rope is measured to be 10 N. What is the tension in the rope when the block is at the bottom of the circle?
>
>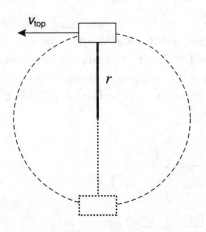

Let's begin by drawing a free-body diagram of the block at the top of the circle and another of the block at the bottom of the circle.

TOP BOTTOM

Next, we write Newton's second law for each diagram. Acceleration is always toward the center of the circle.

$$Mg + T_{\text{top}} = Ma \quad T_{\text{bottom}} - Mg = Ma$$

The acceleration is centripetal, so we can plug in v^2/r for both accelerations.

$$Mg + T_{\text{top}} = Mv^2/r \quad T_{\text{bottom}} - Mg = Mv^2/r$$

At both top and bottom, the speed v and the radius r are the same. So Mv^2/r has to be the same at both the top and bottom, allowing us to set the left side of each equation equal to one another.

$$Mg + T_{\text{top}} = T_{\text{bottom}} - Mg$$

With $M = 2$ kg and $T_{\text{top}} = 10$ N, we solve to get $T_{\text{bottom}} = 50$ N.

Car on a Curve

This next problem is a bit easier than the last one, but it's still good practice.

A car of mass m travels around a flat curve that has a radius of curvature r. What is the necessary coefficient of friction such that the car can round the curve with a velocity v?

Before we draw our free-body diagram, we should consider how friction is acting in this case. Imagine this: what would it be like to round a corner quickly while driving on ice? You would slide off the road, right? Another way to put that is to say that without friction, you would be unable to make the turn. Friction provides the centripetal force necessary to round the corner. Bingo! The force of friction must point in toward the center of the curve.

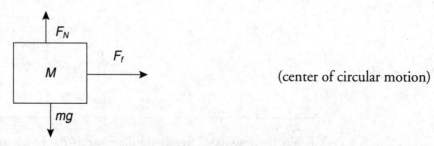

(center of circular motion)

We can now write some equations and solve for μ, the coefficient of friction.

The net force in the horizontal direction is F_f, which can be set equal to mass times (centripetal) acceleration.

$$F_f = ma$$
$$F_f = ma = m\left(\frac{v^2}{r}\right)$$

We also know that $F_f = \mu F_N$. So,

$$\mu F_N = \frac{mv^2}{r}$$

Furthermore, we know that the car is in vertical equilibrium—it is neither flying off the road nor being pushed through it—so $F_N = mg$.

$$\mu mg = \frac{mv^2}{r}$$

Solving for μ we have

$$\mu = \frac{v^2}{gr}$$

Note that this coefficient doesn't depend on mass. Good—if it did, we'd need tires made of different materials depending on how heavy the car is.

Newton's Law of Gravitation

We now shift our focus to gravity. Gravity is an amazing concept—you and the Earth attract each other just because you both have mass!—but at the level tested on the AP exam, it's also a pretty easy concept. In fact, there are only a couple of equations you need to know. The first is for gravitational force:

$$\boxed{F_G = \frac{Gm_1m_2}{r^2}}$$

This equation describes the gravitational force that one object exerts on another object. m_1 is the mass of one of the objects, m_2 is the mass of the other object, r is the distance between the center of mass of each object, and G is called the "Universal Gravitational Constant" and is equal to 6.67×10^{-11} (G does have units—they are $\text{N} \cdot \text{m}^2/\text{kg}^2$—but most problems won't require your knowing them).

> The mass of the Earth, M_E, is 5.97×10^{24} kg. The mass of the sun, M_S, is 1.99×10^{30} kg. The two objects are about 154,000,000 km away from each other. How much gravitational force does Earth exert on the sun?

This is simple plug-and-chug (remember to convert km to m).

$$F_G = \frac{GM_E M_S}{r^2}$$
$$F_G = \frac{6.67 \times 10^{-11}(5.97 \times 10^{24})(1.99 \times 10^{30})}{(1.54 \times 10^{11})^2}$$
$$F_G = 3.3 \times 10^{22} \text{N}$$

Notice that the amount of force that the Earth exerts on the sun is exactly the same as the amount of force the sun exerts on the Earth.

We can combine our knowledge of circular motion and of gravity to solve the following type of problem.

> What is the speed of the Earth as it revolves in orbit around the sun?

The force of gravity exerted by the sun on the Earth is what keeps the Earth in motion—it is the centripetal force acting on the Earth.

$$F_G = M_E a = M_E \left(\frac{v^2}{r} \right)$$

$$\frac{G M_E M_S}{r^2} = \frac{M_E v^2}{r}$$

$$\frac{G M_S}{r} = v^2$$

$$v = \sqrt{\frac{G M_S}{r}}$$

$v = 29,000$ m/s. (Wow, fast . . . that converts to about 14 miles every second—much faster than, say, a school bus.)

Along with the equation for gravitational force, you need to know the equation for gravitational potential energy.

$$PE_G = \frac{G m_1 m_2}{r}$$

We bet you're thinking something like, "Now hold on a minute! You said a while back that an object's gravitational potential energy equals *mgh*. What's going on?"

Good point. An object's gravitational PE equals *mgh* when that object is near the surface of the Earth. But it equals $\frac{G m_1 m_2}{r}$ no matter where that object is.

Similarly, the force of gravity acting on an object equals *mg* (the object's weight) only when that object is near the surface of the Earth.

The force of gravity on an object, however, *always* equals, $\frac{G m_1 m_2}{r^2}$ regardless of location.

› Practice Problems

Multiple Choice:

Questions 1 and 2

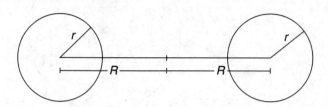

Two stars, each of mass M, form a binary system. The stars orbit about a point a distance R from the center of each star, as shown in the diagram above. The stars themselves each have radius r.

1. What is the force each star exerts on the other?

(A) $G\dfrac{M^2}{(2r+2R)^2}$

(B) $G\dfrac{M^2}{(R+r)^2}$

(C) $G\dfrac{M^2}{R^2}$

(D) $G\dfrac{M^2}{4R^2}$

(E) $G\dfrac{M^2}{2R^2}$

2. In terms of each star's tangential speed v, what is the centripetal acceleration of each star?

(A) $\dfrac{v^2}{2R}$

(B) $\dfrac{v^2}{(r+R)}$

(C) $\dfrac{v^2}{2(r+R)}$

(D) $\dfrac{v^2}{2r}$

(E) $\dfrac{v^2}{R}$

Questions 3 and 4: In the movie *Return of the Jedi*, the Ewoks throw rocks using a circular motion device. A rock is attached to a string. An Ewok whirls the rock in a horizontal circle above his head, then lets go, sending the rock careening into the head of an unsuspecting stormtrooper.

3. What force provides the rock's centripetal acceleration?

(A) The vertical component of the string's tension
(B) The horizontal component of the string's tension
(C) The entire tension of the string
(D) The gravitational force on the rock
(E) The horizontal component of the gravitational force on the rock

4. The Ewok whirls the rock and releases it from a point above his head and to his right. The rock initially goes straight forward. Which of the following describes the subsequent motion of the rock?

(A) It will continue in a straight line forward, while falling due to gravity.

(B) It will continue forward but curve to the right, while falling due to gravity.

(C) It will continue forward but curve to the left, while falling due to gravity.

(D) It will fall straight down to the ground.

(E) It will curve back toward the Ewok and hit him in the head.

5. A Space Shuttle orbits Earth 300 km above the surface. Why can't the Shuttle orbit 10 km above Earth?

(A) The Space Shuttle cannot go fast enough to maintain such an orbit.

(B) Kepler's laws forbid an orbit so close to the surface of the Earth.

(C) Because r appears in the denominator of Newton's law of gravitation, the force of gravity is much larger closer to the Earth; this force is too strong to allow such an orbit.

(D) The closer orbit would likely crash into a large mountain such as Everest because of its elliptical nature.

(E) Much of the Shuttle's kinetic energy would be dissipated as heat in the atmosphere, degrading the orbit.

Free Response:

Center of Rotation

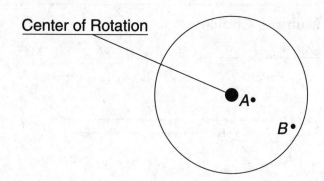

6. Consider two points on a rotating turntable: Point A is very close to the center of rotation, while point B is on the outer rim of the turntable. Both points are shown above. A penny could be placed on the turntable at either point A or point B.

(a) In which case would the speed of the penny be greater, if it were placed at point A, or if it were placed at point B? Explain.

(b) At which point would the penny require the larger centripetal force to remain in place? Justify your answer.

(c) Point B is 0.25 m from the center of rotation. If the coefficient of friction between the penny and the turntable is $\mu = 0.30$, calculate the maximum linear speed the penny can have there and still remain in circular motion.

› Solutions to Practice Problems

1. D—In Newton's law of gravitation,

$$F = G\frac{m_1 m_2}{r^2},$$

the distance used is the distance between the centers of the planets; here that distance is $2R$. But the denominator is squared, so $(2R)^2 = 4R^2$ in the denominator here.

2. E—In the centripetal acceleration equation

$$a_c = \frac{v^2}{r},$$

the distance used is the radius of the circular motion. Here, because the planets orbit around a

point right in between them, this distance is simply R.

3. B—Consider the vertical forces acting on the rock. The rock has weight, so mg acts down. However, because the rock isn't falling down, something must counteract the weight. That something is the vertical component of the rope's tension. The rope must not be perfectly horizontal, then. Because the circle is horizontal, the centripetal force must be horizontal as well. The only horizontal force here is the horizontal component of the tension. (Gravity acts *down*, last we checked, and so cannot have a horizontal component.)

4. A—Once the Ewok lets go, no forces (other than gravity) act on the rock. So, by Newton's first law, the rock continues in a straight line. Of course, the rock still must fall because of gravity. (The Ewok in the movie who got hit in the head forgot to let go of the string.)

5. E—A circular orbit is allowed at any distance from a planet, as long as the satellite moves fast enough. At 300 km above the surface Earth's atmosphere is practically nonexistent. At 10 km, though, the atmospheric friction would quickly cause the shuttle to slow down.

6. (a) Both positions take the same time to make a full revolution. But point B must go farther in that same time, so the penny must have bigger speed at point B.

(b) The coin needs more centripetal force at point B. The centripetal force is equal to mv^2/r. However, the speed itself depends on the radius of the motion, as shown in part (a). The speed of a point at radius r is the circumference divided by the time for one rotation T, $v = 2\pi r/T$. So the net force equation becomes, after algebraic simplification, $F_{net} = 4m\pi^2 r/T^2$. Because both positions take equal times to make a rotation, the coin with the larger distance from the center needs more centripetal force.

(c) The force of friction provides the centripetal force here, and is equal to μ times the normal force. Because the only forces acting vertically are F_N and mg, $F_N = mg$. The net force is equal to mv^2/r, and also to the friction force μmg. Setting these equal and solving for v,

$$v = \sqrt{\mu rg}$$

Plug in the values given ($r = 0.25$ m, $\mu = 0.30$) to get $v = 0.87$ m/s. If the speed is faster than this, then the centripetal force necessary to keep the penny rotating increases, and friction can no longer provide that force.

› Rapid Review

- When an object travels in a circle, its velocity vector is directed tangent to the circle, and its acceleration vector points toward the center of the circle.

- A centripetal force keeps an object traveling in a circle. The centripetal force is simply whatever net force is directed toward the center of the circle in which the object is traveling.

- Newton's law of gravitation states that the gravitational force between two objects is proportional to the mass of the first object multiplied by the mass of the second divided by the square of the distance between them. This also means that the gravitational force felt by one object is the same as the force felt by the second object.

Simple Harmonic Motion

IN THIS CHAPTER

Summary: An object whose position–time graph makes a sine or cosine function is in simple harmonic motion. The period of such motion can be calculated.

Key Ideas

✪ There are two conditions for something to be in simple harmonic motion. Both are equivalent.

1. The object's position–time graph is a sine or cosine graph.
2. The restoring force on the object is proportional to its displacement from equilibrium.

✪ The mass on a spring is the most common example of simple harmonic motion.
✪ The pendulum is in simple harmonic motion for small amplitudes.

Relevant Equations
Period of a mass on a spring:

$$T = 2\pi\sqrt{\frac{m}{k}}$$

Period of a pendulum:

$$T = 2\pi\sqrt{\frac{L}{g}}$$

Relationship between period and frequency:

$$T = \frac{1}{f}$$

What's so simple about simple harmonic motion (SHM)? Well, the name actually refers to a type of movement—regular, back and forth, and tick-tock tick-tock kind of motion. It's simple compared to, say, a system of twenty-five springs and masses and pendulums all tied to one another and waggling about chaotically.

The other reason SHM is simple is that, on the AP exam, there are only a limited number of situations in which you'll encounter it. Which means only a few formulas to memorize, and only a few types of problems to really master.

Amplitude, Period, and Frequency

Simple harmonic motion is the study of oscillations. An **oscillation** is motion of an object that regularly repeats itself over the same path. For example, a pendulum in a grandfather clock undergoes oscillation: it travels back and forth, back and forth, back and forth . . . Another term for oscillation is "periodic motion."

Objects undergo oscillation when they experience a **restoring force**. This is a force that restores an object to the equilibrium position. In the case of a grandfather clock, the pendulum's equilibrium position—the position where it would be if it weren't moving—is when it's hanging straight down. When it's swinging, gravity exerts a restoring force: as the pendulum swings up in its arc, the force of gravity pulls on the pendulum, so that it eventually swings back down and passes through its equilibrium position. Of course, it only remains in its equilibrium position for an instant, and then it swings back up the other way. A restoring force doesn't need to bring an object to rest in its equilibrium position; it just needs to make that object pass through an equilibrium position.

If you look back at the chapter on conservation of energy (Chapter 14), you'll find the equation for the force exerted by a spring, $F = kx$. This force is a restoring force: it tries to pull or push whatever is on the end of the spring back to the spring's equilibrium position. So if the spring is stretched out, the restoring force tries to squish it back in, and if the spring is compressed, the restoring force tries to stretch it back out.[1]

One repetition of periodic motion is called a **cycle**. For the pendulum of a grandfather clock, one cycle is equal to one back-and-forth swing.

The maximum displacement from the equilibrium position during a cycle is the **amplitude**. In Figure 16.1, the equilibrium position is denoted by "0," and the maximum displacement of the object on the end of the spring is denoted by "*A*."

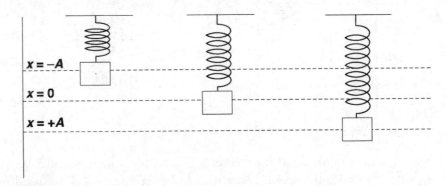

$x = -A$

$x = 0$

$x = +A$

Figure 16.1 Periodic motion of a mass connected to a spring.

---

[1]Some books present this equation as $F = -kx$. The negative sign simply signifies that F is a restoring force.

The time it takes for an object to pass through one cycle is the period, abbreviated "*T*." Going back to the grandfather clock example, the period of the pendulum is the time it takes to go back and forth once: one second. Period is related to frequency, which is the number of cycles per second. The frequency of the pendulum of the grandfather clock is *f* = 1 cycle/s, where "*f*" is the standard abbreviation for frequency; the unit of frequency, the cycle per second, is called a hertz, abbreviated Hz. Period and frequency are related by this equation:

$$T = \frac{1}{f}$$

Vibrating Mass on a Spring

A mass attached to the end of a spring will oscillate in simple harmonic motion. The period of the oscillation is found by this equation:

$$T = 2\pi\sqrt{\frac{m}{k}}$$

In this equation, *m* is the mass of the object on the spring, and *k* is the "spring constant." As far as equations go, this is one of the more difficult ones to memorize, but once you have committed it to memory, it becomes very simple to use.

A block with a mass of 10 kg is placed on the end of a spring that is hung from the ceiling. When the block is attached to the spring, the spring is stretched out 20 cm from its rest position. The block is then pulled down an additional 5 cm and released. What is the block's period of oscillation, and what is the speed of the block when it passes through its rest position?

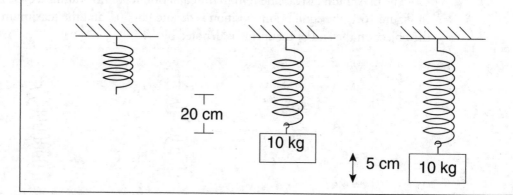

Let's think about how to solve this problem methodically. We need to find two values, a period and a speed. Period should be pretty easy—all we need to know is the mass of the block (which we're given) and the spring constant, and then we can plug into the formula. What about the speed? That's going to be a conservation of energy problem—potential energy in the stretched-out spring gets converted to kinetic energy—and here again, to calculate the potential energy, we need to know the spring constant. So let's start by calculating that.

First, we draw our free-body diagram of the block.

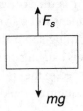

We'll call "up" the positive direction. Before the mass is oscillating, the block is in equilibrium, so we can set F_s equal to mg. (Remember to convert centimeters to meters!)

$$kx = mg$$
$$k\,(0.20\text{ m}) = (10\text{ kg})(10\text{ m/s}^2)$$
$$k = 500\text{ N/m}$$

Now that we have solved for k, we can go on to the rest of the problem. The period of oscillation can be found be plugging into our formula.

$$T = 2\pi\sqrt{\frac{m}{k}}$$

$$T = 2\pi\sqrt{\frac{10\text{ kg}}{500\text{ N/m}}}$$

$$T = 0.89\text{ s}$$

To compute the velocity at the equilibrium position, we can now use conservation of energy.

$$\text{KE}_a + \text{PE}_a = \text{KE}_b + \text{PE}_b$$

When dealing with a vertical spring, it is best to define the rest position as $x = 0$ in the equation for potential energy of the spring. If we do this, then gravitational potential energy can be ignored. Yes, gravity still acts on the mass, and the mass changes gravitational potential energy. So what we're really doing is taking gravity into account in the spring potential energy formula by redefining the $x = 0$ position, where the spring is stretched out, as the resting spot rather than where the spring is unstretched.

In the equation above, we have used a subscript "a" to represent values when the spring is stretched out the extra 5 cm, and "b" to represent values at the rest position.

When the spring is stretched out the extra 5 cm, the block has no kinetic energy because it is being held in place. So, the KE term on the left side of the equation will equal 0. At this point, all of the block's energy is entirely in the form of potential energy. (The equation for the PE of a spring is $\frac{1}{2}kx^2$, remember?) And at the equilibrium position, the block's energy will be entirely in the form of kinetic energy. Solving, we have

$$\tfrac{1}{2}mv_a^2 + \tfrac{1}{2}kx_a^2 = \tfrac{1}{2}mv_b^2 + \tfrac{1}{2}kx_b^2$$
$$0 + \tfrac{1}{2}\,(500\text{ N/m})(0.05\text{ m})^2 = \tfrac{1}{2}\,(10\text{ kg})v^2 + 0$$
$$v = 0.35\text{ m/s}$$

Simple Pendulums

Problems that involve simple pendulums—in other words, basic, run-of-the-mill, grandfather clock–style pendulums—are actually really similar to problems that involve springs. For example, the formula for the period of a simple pendulum is this:

$$T = 2\pi\sqrt{\frac{L}{g}}$$

Looks kind of like the period of a mass on a spring, right? In this equation, L is the length of the pendulum, and g is the acceleration attributable to gravity (about 10 m/s^2). Of course, if your pendulum happens to be swinging on another planet, g will have a different value.[2]

One interesting thing about this equation: The period of a pendulum does not depend on the mass of whatever is hanging on the end of the pendulum. So if you had a pendulum of length L with a peanut attached to the end, and another pendulum of length L with an elephant attached to the end, both pendulums would have the same period in the absence of air resistance.

A string with a bowling ball tied to its end is attached to the ceiling. The string is pulled back such that it makes a 10° angle with the vertical, and it is then released. When the bowling ball reaches its lowest point, it has a speed of 2 m/s. What is the frequency of this bowling-ball pendulum?

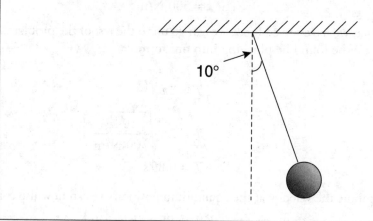

To calculate the period of this pendulum, we must know the length of the string. We can calculate this using conservation of energy. Then, we'll convert the period to a frequency.

Before the string is released, all of the bowling ball's energy is in the form of gravitational PE. If we define the zero of potential to be at the ball's lowest point, then at that point all the bowling ball's energy is in the form of KE. We will use a subscript "a" to represent values before the bowling ball is released and "b" to represent values when the bowling ball is at its lowest point.

$$KE_a + PE_a = KE_b + PE_b$$

$$\tfrac{1}{2}mv_a^2 + mgh_a = \tfrac{1}{2}mv_b^2 + mgh_b$$

$$0 + mgh_a = \tfrac{1}{2}mv_b^2 + 0$$

The height of the bowling ball before it is released, h_a, can be calculated using trigonometry.

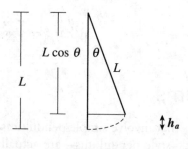

$$h_a = L - L\cos\theta$$

So, getting back to our previous equation, we have

$$0 + mg(L - L\cos\theta) = \tfrac{1}{2}mv_b^2 + 0$$

---

[2]But even if you did travel to another planet, do you really think you would remember to pack your pendulum?

We know θ and we know v_b, so we can solve for L.

$$L - L \cos \theta = (1/g) \,{}^1\!/_2 v_b^2$$

$$L (1 - \cos \theta) = (1/g) \,{}^1\!/_2 v_b^2$$

$$L = 13.2 \text{ m}$$

Now that we know L, we can find the frequency.

$$T = 2\pi \sqrt{\frac{L}{g}}$$

$$T = 2\pi \sqrt{\frac{13.2}{100}}$$

$T = 7.2$ s; frequency is $1/T$, or 0.14 Hz

› Practice Problems

1. A basketball player dribbles the ball so that it bounces regularly, twice per second. Is this ball in simple harmonic motion? Explain.

Multiple Choice:

2. A pendulum has a period of 5 seconds on Earth. On Jupiter, where $g \sim 30$ m/s², the period of this pendulum would be closest to

(A) 1 s
(B) 3 s
(C) 5 s
(D) 8 s
(E) 15 s

3. A pendulum and a mass on a spring are designed to vibrate with the same period T. These devices are taken onto the Space Shuttle in orbit. What is the period of each on the Space Shuttle?

pendulum	mass on a spring
(A) will not oscillate	T
(B) T	will not oscillate
(C) will not oscillate	will not oscillate
(D) $\sqrt{2} \cdot T$	T
(E) T	$\sqrt{2} \cdot T$

4. A mass on a spring has a frequency of 2.5 Hz and an amplitude of 0.05 m. In one complete period, what distance does the mass traverse? (This question asks for the actual distance, not the displacement.)

(A) 0.05 cm
(B) 0.01 cm
(C) 20 cm
(D) 10 cm
(E) 5 cm

5. Increasing which of the following will increase the period of a simple pendulum?

  I. the length of the string
 II. the local gravitational field
III. the mass attached to the string

(A) I only
(B) II only
(C) III only
(D) I and II only
(E) I, II, and III

Free Response:

6. A mass m is attached to a horizontal spring of spring constant k. The spring oscillates in simple harmonic motion with amplitude A. Answer the following in terms of A.

(a) At what displacement from equilibrium is the speed half of the maximum value?

(b) At what displacement from equilibrium is the potential energy half of the maximum value?

(c) When is the mass farther from its equilibrium position, when its speed is half maximum, or when its potential energy is half maximum?

› Solutions to Practice Problems

1. The ball is *not* in simple harmonic motion. An object in SHM experiences a force that pushes toward the center of the motion, pushing harder the farther the object is from the center; and, an object in SHM oscillates smoothly with a sinusoidal position–time graph. The basketball experiences only the gravitational force, except for the brief time that it's in contact with the ground. Its position–time graph has sharp peaks when it hits the ground.

2. **B**—The period of a pendulum is

$$T = 2\pi\sqrt{\frac{L}{g}}$$

All that is changed by going to Jupiter is g, which is multiplied by 3. g is in the denominator and under a square root, so the period on Jupiter will be *reduced* by a factor of $\sqrt{3}$. So the original 5-second period is cut by a bit less than half, to about 3 seconds.

3. **A**—The restoring force that causes a pendulum to vibrate is gravity. Because things float in the Space Shuttle rather than fall to the floor, the pendulum will not oscillate at all. However, the restoring force that causes a *spring* to vibrate is the spring force itself, which does not depend on gravity. The period of a mass on a spring also depends on mass, which is unchanged in the Space Shuttle, so the period of vibration is unchanged as well.

4. **C**—The amplitude of an object in SHM is the distance from equilibrium to the maximum displacement. In one full period, the mass traverses this distance four times: starting from max displacement, the mass goes down to the equilibrium position, down again to the max displacement on the opposite side, back to the equilibrium position, and back to where it started from. This is 4 amplitudes, or 0.20 m, or 20 cm.

5. **A**—The period of a pendulum is

$$T = 2\pi\sqrt{\frac{L}{g}}$$

Because L, the length of the string, is in the numerator, increasing L increases the period. Increasing g will actually *decrease* the period because g is in the denominator; increasing the mass on the pendulum

has no effect because mass does not appear in the equation for period.

6. (a) The maximum speed of the mass is at the equilibrium position, where PE = 0, so all energy is kinetic. The maximum potential energy is at the maximum displacement A, because there the mass is at rest briefly and so has no KE. At the equilibrium position all of the PE has been converted to KE, so

$$\tfrac{1}{2}kA^2 = \tfrac{1}{2}mv^2_{max}$$

Solving for v_{max}, it is found that

$$v_{max} = A\sqrt{\frac{k}{m}}$$

Now that we have a formula for the maximum speed, we can solve the problem. Call the spot where the speed is half-maximum position 2. Use conservation of energy to equate the energy of the maximum displacement and position 2:

$$\tfrac{1}{2}kA^2 = \tfrac{1}{2}mv_2^2 + \tfrac{1}{2}kx_2^2$$

The speed v_2 is half of the maximum speed we found earlier, or $\tfrac{1}{2}\left(A\sqrt{\dfrac{k}{m}}\right)$. Plug that in and solve for x_2:

$$\tfrac{1}{2}kA^2 = \tfrac{1}{2}m\left(\tfrac{1}{2}A\sqrt{\frac{k}{m}}\right)^2 + \tfrac{1}{2}kx_2^2$$

$$\tfrac{1}{2}kA^2 = \tfrac{1}{4}mA^2\frac{k}{m} + \tfrac{1}{2}kx_2^2$$

The m's and the k's cancel. The result is $x_2 = A\sqrt{\dfrac{3}{4}}$, or about 86% of the amplitude.

(b) The total energy is $\tfrac{1}{2}kA^2$. At some position x, the potential energy will be $\tfrac{1}{2}$ of its maximum value. At that point, $\tfrac{1}{2}kx^2 = \tfrac{1}{2}(\tfrac{1}{2}kA^2)$. Canceling and solving for x, it is found that

$$x = \frac{1}{\sqrt{2}}A$$

This works out to about 70% of the maximum amplitude.

(c) Since we solved in terms of A, we can just look at our answers to (a) and (b). The velocity is half-maximum at $\sqrt{\frac{3}{4}}$, or 86%, of A; the

potential energy is half-maximum at $\frac{1}{\sqrt{2}}$ or 71%, of A. Therefore, the mass is farther from equilibrium when velocity is half-maximum.

› Rapid Review

- An oscillation is motion that regularly repeats itself over the same path. Oscillating objects are acted on by a restoring force.

- One repetition of periodic motion is called a cycle. The maximum displacement of an oscillating object during a cycle is the object's amplitude. The time it takes for an object to go through a cycle is the period of oscillation.

- Period is related to frequency: $T = 1/f$, and $f = 1/T$.

- When solving problems that involve springs or simple pendulums, be on the lookout for ways to apply conservation of energy. Not every simple harmonic motion problem will require you to use conservation of energy, but many will.

CHAPTER 17

Thermodynamics

IN THIS CHAPTER

Summary: Energy transfer in gases can do measurable work on large objects. Thermodynamics is the study of heat energy and its consequences.

Key Ideas

✪ Internal energy is the energy possessed by a substance due to the vibration of molecules. A substance's internal energy depends on its temperature.

✪ Heat is a transfer of thermal energy.

✪ When an object's temperature increases, it expands.

✪ The ideal gas law relates three important thermodynamics variables: pressure, temperature, and volume.

✪ Kinetic theory explains how the motion of many individual molecules in a gas leads to measurable properties such as temperature.

✪ The first law of thermodynamics is a statement of conservation of energy. It is usually applied when a gas's state is represented on a Pressure vs. Volume (PV) diagram. From a PV diagram:

(a) Internal energy can be determined using the ideal gas law and PV values from the diagram's axes.

(b) Work done on or by the gas can be determined from the area under the graph.

(c) Heat added or removed from the gas can NOT be determined directly from the diagram, but rather only from the first law of thermodynamics.

✪ The efficiency of an engine is defined as work output divided by energy input. The efficiency of an ideal heat engine is NOT 100%, but rather is given by the Carnot formula.

Relevant Equations
Thermal length expansion:

$$\Delta L = \alpha L_0 \Delta T$$

Ideal gas law:

$$PV = nRT$$

Internal energy of an ideal gas:

$$U = \tfrac{3}{2}\, nRT \left(= \tfrac{3}{2} PV\right)$$

RMS speed of a gas molecule:

$$v_{\text{rms}} = \sqrt{\frac{3k_B T}{m}}$$

First law of thermodynamics:

$$\Delta U = Q + W$$

Efficiency of any heat engine:

$$e = \frac{W}{Q_H}$$

Efficiency of an ideal (Carnot) engine:

$$e = \frac{T_H - T_C}{T_H}$$

Rate of heat transfer:

$$H = \frac{kA\Delta T}{L}$$

Thermodynamics, as the name implies, is the study of what heat is and how it gets transferred. It's a very different topic from most of the other subjects in this book because the systems we focus on are, for the most part, molecular. This doesn't mean that such things as blocks and inclined planes won't show up in a thermodynamics problem—in fact, we've included a problem in this chapter that involves a sliding block. But molecules and gases are at the heart of thermodynamics, so this chapter includes a lot of material introducing you to them.

Heat, Temperature, and Power

Heat, represented by the variable Q, is a type of energy that can be transferred from one body to another. As you might expect, heat is measured in joules, because it is a form of energy. Notice that all energy quantities (KE, PE, Work) have units of joules. Heat is no different.

Be careful with phraseology here. Energy must be *transferred* in order to be called heat. So, heat can be gained or lost, but not possessed. It is incorrect to say, "a gas has 3000 J of heat."

> **Internal Energy:** The sum of the energies of all of the molecules in a substance

Internal energy, represented by the variable U, is also measured in joules, again because it is a form of energy. It *is* correct to say, "a gas has 3000 J of internal energy."

> **Temperature:** Related to the average kinetic energy per molecule of a substance

Temperature is measured in kelvins or degrees Celsius. There is no such thing as "a degree kelvin"; it's just "a kelvin." To convert from kelvins to degrees Celsius or *vice versa*, use the following formula:

$$\text{Temp}_{\text{Celsius}} = \text{Temp}_{\text{Kelvin}} - 273$$

So, if room temperature is about 23°C, then room temperature is also about 296 K.

It is important to note that *two bodies that have the same temperature do not necessarily contain the same amount of internal energy.* For example, let's say you have a huge hunk of metal at 20°C and a tiny speck of the same type of metal, also at 20°C. The huge hunk contains a lot more internal energy—it has a lot more molecules moving around—than the tiny speck. By the same logic, you could have two items with very different temperatures that contain exactly the same amount of internal energy.

> **Power:** Work per time

Power is measured in joules/second, or watts:

$$1 \text{ W} = 1 \text{ J/s}$$

Power is not an idea limited to thermodynamics. In fact, we often talk about power in the context of mechanics. We might ask, for example, "How much power is needed to raise a block of mass 2 kg a distance of 1 meter in 4 seconds?"[1] But discussing power now allows us to bridge mechanics and thermodynamics . . . with this problem.

> In a warehouse, a 200-kg container breaks loose from a forklift. The container slides along the floor at a velocity of 5 m/s. The force of friction between the block and the floor produces thermal energy (heat) at a rate of 10 W. How far does the block travel before coming to rest?

The force of friction is converting the block's kinetic energy into heat—this is why friction slows objects down and why your hands get hot when you rub them together. We know the rate at which the block's KE is being converted to heat, so if we calculate how much KE it had to begin with, we can find how long it takes to come to rest.

$$\text{KE}_i = \tfrac{1}{2} \, mv^2 = \tfrac{1}{2} (200 \text{ kg})(5 \text{ m/s})^2 = 2500 \text{ J}$$

$$\text{KE}_f = 0$$

$$\frac{2500 \text{J}}{10 \frac{\text{J}}{\text{s}}} = 250 \text{ s}$$

---

[1] 5 watts—much less power than put out by a light bulb.

Now that we know the length of time the block was sliding, we can determine the distance it traveled using kinematics. You know how to do this well already, so we won't take you through it step-by-step. Just remember to fill in your table of variables and go from there. The answer is 625 m, or more than a quarter mile. . . this must be most of the way across the warehouse!

Thermal Expansion

Things get slightly bigger when they heat up. We call this phenomenon **thermal expansion**. This is why a thermometer works. The mercury in the thermometer expands when it gets hotter, so the level of mercury rises. When it gets colder, the mercury shrinks and the level of mercury drops. The equation describing thermal expansion follows:

$$\Delta L = \alpha L_0 \Delta T$$

This formula describes what is called linear expansion. It says that the change in an object's length, ΔL, is equal to some constant, α, multiplied by the object's original length, L_0, multiplied by the change in the object's temperature.

The constant α is determined experimentally: it is different for every material. Metals tend to have high values for α, meaning that they expand a lot when heated. Such insulators as plastic tend to have low values for α. But on the AP exam, you will never be expected to have memorized the value of α for any particular material. Here's an example involving thermal expansion.

> You need to slide an aluminum ring onto a rod. At room temperature (about 20°C), the internal diameter of the ring is 50.0 mm and the diameter of the rod is 50.1 mm. (a) Should you heat or cool the ring to make it fit onto the rod? (b) If the coefficient of linear expansion for aluminum is $2.5 \times 10^{-5}\,°\text{C}^{-1}$, at what temperature will the ring just barely fit onto the rod?

(a) You should **heat the ring**. You might initially think that because you want the inner diameter of the ring to expand, you want the aluminum itself to shrink. However, when a material expands, it expands in all directions—thus, the hole will expand by the same amount as the surrounding metal when heated.

(b) In the length expansion equation, ΔL is the amount by which we need the diameter of the ring to increase, 0.1 mm. Remember to convert to meters before solving.

$$0.0001 \text{ m} = (2.5 \times 10^{-5}\,°\text{C}^{-1})(.050 \text{ m})(\Delta T)$$

Solving for ΔT, we find that we need to increase the ring's temperature by 80°C. So the final temperature of the ring is (20 + 80) = 100°C.

Two more notes: (1) Thermal expansion equations assume that length changes will be small compared to the size of the object in question. For example, the object above only expanded by a few tenths of a percent of its original size. Objects will not double in size due to thermal expansion. (2) If the standard unit of temperature is the kelvin, why did we use degrees Celsius in the example? It is true that if an equation includes the variable T for temperature, you *must* plug in temperature in kelvins. The ideal gas law is one such equation. However, the equation for thermal expansion includes a **change** in temperature, not an absolute temperature. A change of one degree Celsius is equivalent to a change of one kelvin; so if an equation needs only a temperature change, then either unit is acceptable.

Rate of Heat Transfer

Heat will flow across a material when there is a temperature difference between two positions on that material. Heat will flow until the material has the same temperature everywhere, when the material is said to have reached **thermal equilibrium**. The bigger the temperature difference, the larger the heat transfer rate. You could probably guess at the other properties of the material that affect the rate of heat transfer:

- the nature of the material (metals carry heat better than non-metals)
- the cross-sectional area of the substance (more area makes it easier for heat to flow)
- the distance across the material between the two positions with a temperature difference (the longer this distance, the slower the heat transfer)

This can be summed up in the equation for the heat transfer rate H:

$$H = \frac{kA\Delta T}{L}$$

where k is the thermal conductivity of the substance, A is the cross-sectional area, ΔT is the temperature difference, and L is the length of the material.

Ideal Gas Law

This is probably one of your favorite laws, because you used it so much in chemistry class. It can be written two different ways; our preference is

$$\boxed{PV = nRT}$$

P is the pressure of the gas, V is the volume of the gas, n is the number of moles of gas you're dealing with, R is a constant ($R = 8.31$ J/mol·K), and T is the temperature of the gas in kelvins. And in case you were wondering, the units of pressure are N/m², also known as pascals, and the units of volume are m³. (Yes, there are other units for pressure and volume, but don't use them unless you want to be really confused.)

The ideal gas law can also be written as

$$\boxed{PV = Nk_BT}$$

In this version, N is the number of *molecules* of gas, and k_B is called Boltzmann's constant ($k_B = 1.38 \times 10^{-23}$ J/K).

If you ever need to convert moles to molecules, remember that one mole contains 6.02×10^{23} molecules. This conversion is on the constant sheet.

The rigid frame of a dirigible holds 5000 m³ of hydrogen. On a warm 20°C day, the hydrogen is at atmospheric pressure. What is the pressure of the hydrogen if the temperature drops to 0°C?

The best way to approach an ideal gas problem is to ask: "What is constant?" In this case, we have a *rigid* container, meaning that volume of the hydrogen cannot change. Also, unless there's a hole in the dirigible somewhere, the number of molecules N must stay constant. So the only variables that change are pressure and temperature.

Start by putting the constants on the right-hand side of the ideal gas equation:

$$\frac{P}{T} = \frac{Nk_B}{V}$$

Because the right side doesn't change, we know that the pressure divided by the temperature always stays the same for this dirigible:

$$\frac{10^5\,\text{N}/\text{m}^2}{293\text{K}} = \frac{P}{273\text{K}}$$

Here atmospheric pressure, as given on the constant sheet, is plugged in (note that temperature must always be in units of kelvins in this equation). Now just solve for P. The answer is 93,000 N/m², or just a bit less than normal atmospheric pressure.

Kinetic Theory of Gases

Kinetic theory makes several sweeping assumptions about the behavior of molecules in a gas. These assumptions are used to make predictions about the macroscopic behavior of the gas. Don't memorize these assumptions. Rather, picture a bunch of randomly bouncing molecules in your mind, and see how these assumptions do actually describe molecular behavior.

- Molecules move in continuous, random motion.
- There are an exceedingly large number of molecules in any container of gas.
- The separation between individual molecules is large.
- Molecules do not act on one another at a distance; that is, they do not exert electrical or gravitational forces on other molecules.
- All collisions between molecules, or between a molecule and the walls of a container, are elastic (i.e., kinetic energy is not lost in these collisions).

What do you need to know about kinetic theory? Well, you need to understand the above assumptions. And two results of kinetic theory are important. These are derived using Newtonian mechanics and the assumptions. Don't worry for now about the derivation, just learn the equations and what they mean:

1. The relationship between the internal energy of a gas and its temperature is

$$U = \tfrac{3}{2}\,nRT$$

This equation allows you to find the temperature of a gas given its internal energy, or vice versa.

2. The root-mean-square (rms) speed of gas molecules is

$$v_{\text{rms}} = \sqrt{\frac{3k_B T}{m}}$$

Here, T is the temperature of the gas (always in kelvins!) and m is the mass of one molecule of the gas (always in kilograms). Think of the rms speed as a sort of average speed of molecules in a gas. For most gases, this speed will be in the hundreds or thousands of meters per second.

First Law of Thermodynamics

Before you start this section, be sure you understand the difference between heat and internal energy, as described several pages ago. Furthermore, remember that mechanical work is done whenever a force is applied through a distance.

The first law of thermodynamics is simply a statement of conservation of energy.

$$\Delta U = Q + W$$

U is the internal energy of the gas in question. Q is the heat added to the gas, and W is the work done on the gas when the gas is compressed.

The signs of each term in the first law are extremely important. If heat is added to the gas, Q is positive, and if heat is taken from the gas, Q is negative. If the gas's internal energy (and, thus, its temperature) goes up, then ΔU is positive; if the gas's temperature goes down, then ΔU is negative.

The sign of W is more difficult to conceptualize. Positive W represents work done *on* a gas. However, if a gas's volume expands, then work was not done *on* the gas, work was done *by* the gas. In the case of work done *by* a gas, W is given a negative sign.

> **Exam tip from an AP Physics veteran:**
> To memorize all these sign conventions, repeat the following phrase out loud, at least twice a day, every day between now and the AP exam—"Heat added to plus work done on." Say it with a rhythm: baa-bi-bi-baa-baa-baa-baa-baa. If you repeat this phrase enough, you'll always remember that heat added to a gas, and work done on a gas, are positive terms.
>
> —*Zack, college senior and engineer*

Here's a sample problem.

> A gas is kept in a cylinder that can be compressed by pushing down on a piston. You add 2500 J of heat to the system, and then you push the piston 1.0 m down with a constant force of 1800 N. What is the change in the gas's internal energy?

Define the variables in the first law, being careful of signs: Here $Q = +2500$ J. We're pushing the piston down; therefore, work is done *on* the gas, and W is positive. We remember that work = force × distance; so, the work done on the gas here is 1800 N·1 m, or +1800 J. Now plug into the first law of thermodynamics: the answer is +4300 J.

PV Diagrams

Imagine a gas in a cylinder with a moveable piston on top, as shown in Figure 17.1.

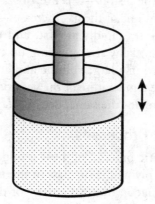

Figure 17.1   Gas in a cylinder with a moveable piston.

When the gas expands, the piston moves up, and work is done *by* the gas. When the piston compresses, work is done *on* the gas. The point of a **heat engine** is to add heat to a gas, making the gas expand—in other words, using heat to do some work by moving the piston up.

To visualize this process mathematically, we use a graph of pressure vs. volume, called a ***PV* diagram**. On a *PV* diagram, the *x*-axis shows the volume of the gas, and the *y*-axis shows the pressure of the gas at a particular volume. For the AP exam, you're expected to be familiar with the *PV* diagrams for four types of processes:

- **Isothermal:** The gas is held at a constant temperature.
- **Adiabatic:** No heat flows into or out of the gas.
- **Isobaric:** The gas is held at a constant pressure.
- **Isochoric:** The gas is held at a constant volume.

Not all processes on a *PV* diagram fall into one of these four categories, but many do. We look at each of these processes individually.

Isothermal Processes

Look at the ideal gas law: $PV = Nk_B T$. In virtually all of the processes you'll deal with, no gas will be allowed to escape, so N will be constant. Boltzmann's constant is, of course, always constant. And, in an isothermal process, T must be constant. So the entire right side of the equation remains constant during an isothermal process. What this means is that the product of P and V must also be constant. A *PV* diagram for an isothermal process is shown in Figure 17.2.

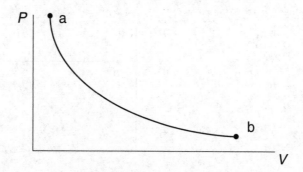

Figure 17.2   *PV* diagram of an isothermal process.

What this diagram shows is the set of all *P–V* values taken on by a gas as it goes from point "a" to point "b." At point "a," the gas is held at high pressure but low volume. We can imagine that there's a lot of gas squished into a tiny cylinder, and the gas is trying its hardest to push up on the piston—it is exerting a lot of pressure.

As the volume of the gas increases—that is, as the piston is pushed up—the pressure of the gas decreases rapidly at first, and then slower as the volume of the gas continues to increase. Eventually, after pushing up the piston enough, the gas reaches point "b," where it is contained in a large volume and at a low pressure.

Since the temperature of the gas does not change during this process, the internal energy of the gas remains the same, because $U = (3/2)Nk_b T$. *For an isothermal process,* $\Delta U = 0$.

Adiabatic Processes

In an adiabatic process, no heat is added to or removed from the system. In other words, $Q = 0$. By applying the first law of thermodynamics, we conclude that $\Delta U = W$. This means that the change in internal energy of the gas equals the work done ON the system. In an adiabatic process, you can increase the temperature of the gas by pushing down on the piston, or the temperature of the gas can decrease if the gas itself pushes up on the piston. The *PV* diagram for an adiabatic process looks similar, but not identical, to the diagram for an isothermal process—note that when a gas is expanded adiabatically, as from "a" to "b" in Figure 17.3, the final state is on a lower isotherm. (An isotherm is any curve on a *PV* diagram for which the temperature is constant.) In other words, in an adiabatic expansion, the temperature of the gas decreases.

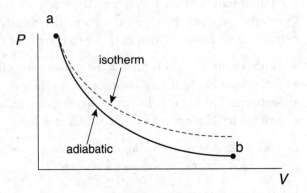

Figure 17.3 *PV* **diagram of an adiabatic process.**

Isobaric Processes

This one's pretty easy: the pressure stays constant. So a *PV* diagram for an isobaric process looks like the graph in Figure 17.4.

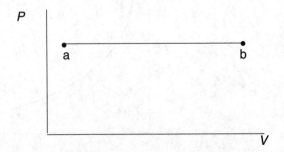

Figure 17.4 *PV* **diagram of an isobaric process.**

In an isobaric process, the pressure of the gas is the same no matter what the volume of the gas is. Using $PV = Nk_BT$, we can see that, if we increase V but not P, then the left side of the equation increases. So the right side must also: the temperature must increase. Therefore, in an isobaric process, the temperature of the gas increases as the volume increases, but the pressure doesn't change.

Isochoric Processes

Even easier: the volume remains constant, as shown in Figure 17.5.

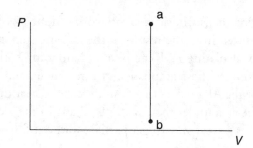

Figure 17.5 *PV* **diagram of an isochoric process.**

As with an isobaric process, the temperature of a gas undergoing an isochoric process must change as the gas goes from point "a" to point "b." The temperature of the gas must increase as the pressure increases, and it must decrease as the pressure of the gas decreases.

Using *PV* Diagrams

Working with *PV* diagrams can seem rather complicated, but, fortunately, you'll rarely come upon a complicated *PV* diagram on the AP exam. We have two tips for solving problems using *PV* diagrams: if you follow them, you should be set for just about anything the AP could throw at you.

TIP #1: Work done on or by a gas = area under the curve on a *PV* diagram.

For an isobaric process, calculating area under the curve is easy: just draw a line down from point "a" to the horizontal axis and another line from point "b" to the horizontal axis, and the area of the rectangle formed is equal to $P\Delta V$ (where ΔV is the change in the volume of the gas). The equation sheet for the AP exam adds a negative sign, indicating that work done *by* the gas (i.e., when the volume expands) is defined as negative: $W = -P\Delta V$. It's even easier for an isochoric process: $-P\Delta V = 0$.

This tip comes in handy for a problem such as the following.

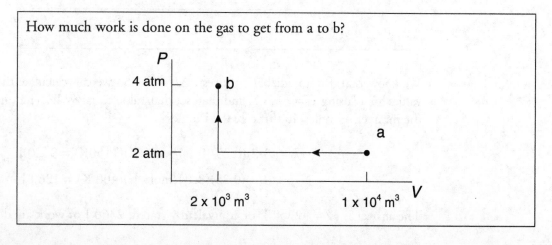

We have two steps here. The first is isobaric, the second is isochoric. During the isobaric step, the work done by the gas is

$$W = -P\Delta V = -(2 \text{ atm})(2000–10{,}000)\text{m}^3$$

Careful—you always must be using standard units so your answer will be in joules!

$$W = -(2 \times 10^5 \text{ N/m}^2)(-8000 \text{ m}^3) = +1.6 \times 10^9 \text{ J}$$

During the isochoric step, no work is done by or on the gas. So the total work done on the gas, taken from the results of the isobaric step, is 1.6×10^9 J.

Calculating $P\Delta V$ can be a bit harder for isothermal and adiabatic processes, because the area under these graphs is not a rectangle, and $-P\Delta V$ does not apply. But if you want to show the AP graders that you understand that the work done by the gas in question is equal to the area under the curve, just shade in the area under the curve and label it accordingly. This is a great way to get partial—or even full—credit.

TIP #2: Remember that $U = \frac{3}{2} nRT$.

Internal energy is directly related to temperature. So, when a process starts and ends at the same point on a *PV* diagram, or even on the same isotherm, $\Delta U = 0$.

This tip is particularly useful for adiabatic processes, because, in those cases, $\Delta U = W$. So being able to calculate ΔU allows you to find the work done on, or by, the system easily. We'll provide an example problem.

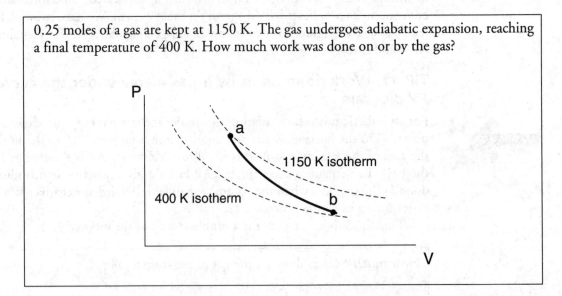

0.25 moles of a gas are kept at 1150 K. The gas undergoes adiabatic expansion, reaching a final temperature of 400 K. How much work was done on or by the gas?

We know that, for an adiabatic process, $\Delta U = W$. So we can calculate the final and initial values for U using $U = \frac{3}{2} nRT$, and then set that value equal to W. The variable n represents the number of moles, in this case 0.25 moles

$$U_a = (3/2)(0.25)(8.1 \text{ J/mol} \cdot \text{K})(1150\text{K}) = 3620 \text{ J}$$

$$U_b = (3/2)(0.25)(8.1 \text{ J/mol} \cdot \text{K})(400 \text{ K}) = 1260 \text{ J}$$

The answer is $W = -2360$ J, or equivalently stated, 2360 J of work are done *by* the gas.

Heat Engines and the Second Law of Thermodynamics

The second law of thermodynamics is less mathematical than the first law, and it is also probably more powerful. One of many ways to state this law says.

> Heat flows naturally from a hot object to a cold object.
> Heat does NOT flow naturally from cold to hot.

Pretty simple, no? Heat engines and their cousins, refrigerators, work by this principle. Figure 17.6 shows what a heat engine looks like to a physicist.

In a heat engine, you start with something hot. Heat must flow from hot stuff to cold stuff . . . some of this flowing heat is used to do work. The rest of the input heat flows into the cold stuff as exhaust. A car engine is a heat engine: you start with hot gas in a cylinder, that hot gas expands to push a piston (it does work), and you're left with colder[2] exhaust gas.

We define the efficiency of a heat engine by the ratio of the useful work done by the engine divided by the amount of heat that was put in. This is stated by the following equation:

$$e = \frac{W}{Q_H}$$

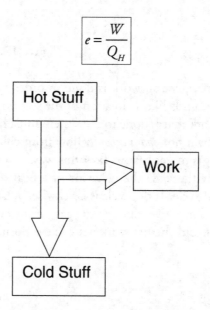

Figure 17.6 A heat engine.

The efficiency, e, of an engine is always less than 1. If the efficiency of an engine were equal to 1, then all the heat in the hot stuff would be turned into work, and you'd be left with material at absolute zero. That's tough to do, so we'll settle for an efficiency less than 1.

Since all heat engines operate at less than 100% efficiency, it is useful to calculate what the maximum possible efficiency might be, assuming the engine is as well made as possible to eliminate energy losses to friction, insufficient insulation, and other engineering concerns. It is found that this ideal efficiency for a heat engine depends only on the temperatures of the hot and cold reservoirs.

$$e_{ideal} = \frac{T_H - T_C}{T_H}$$

---

[2]Okay, well, not cold by normal standards—exhaust gasses can be hotter than boiling water—but colder than the gas in the combustion chamber.

T_H is the temperature of the hot stuff you start with, and T_C is the temperature of the colder exhaust. These temperatures need to be measured in kelvins.[3]

Just because a heat engine is "ideal" doesn't mean its efficiency is 1. The same argument we used above still applies: efficiency is always less than 1. And real, nonideal engines operate at only a fraction of their ideal efficiency.

A refrigerator is like the backward cousin of a heat engine. Physicists draw a refrigerator as shown in Figure 17.7.[4]

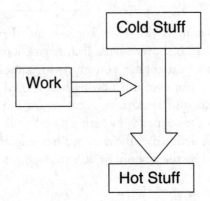

Figure 17.7   A refrigerator.

In a refrigerator, work is done in order to allow heat to flow from cold stuff to hot stuff. This sounds like a violation of the second law of thermodynamics. It isn't, though, because the work that is done to the system must have been the product of some sort of heat engine. Heat will not *spontaneously* flow from cold stuff to hot stuff; it is necessary to do net work on a gas to get heat to flow this way.

By the way: the behavior of a heat engine can be represented on a *PV* diagram by a closed cycle. If the cycle is clockwise, you're looking at a heat engine, and the net work per cycle is negative (done *by* the gas); if the cycle is counterclockwise, you're looking at a refrigerator, and the net work per cycle is positive (done *on* the gas).

Entropy

Entropy is a measure of disorder. A neat, organized room has low entropy . . . the same room after your three-year-old brother plays "tornado!" in it has higher entropy.

There are quantitative measures of entropy, but these are not important for the AP exam. What you do need to know is that entropy relates directly to the second law of thermodynamics. In fact, we can restate the second law of thermodynamics as follows.

> The entropy of a system cannot decrease unless work is done on that system.

This means that the universe moves from order to disorder, not the other way around. For example: A ball of putty can fall from a height and splat on the floor, converting potential energy to kinetic energy and then to heat—the ball warms up a bit. Now, while it is not against the laws of conservation of energy for the putty to convert some of that heat back into

---

[3]The ideal efficiency is often called the "Carnot" efficiency after the 19th century engineer Sadi Carnot. A "Carnot engine" is a putative engine that operates at the maximum possible efficiency (which is not 100%!).

[4]Personally, we don't think this looks anything like a refrigerator. Where's the ice cube maker? Where are the amusing magnets?

kinetic energy and fly back up to where it started from. Does this ever happen? Um, no. The molecules in the putty after hitting the ground were put in a more disordered state. The second law of thermodynamics does not allow the putty's molecules to become more ordered.

Or, a glass can fall off a shelf and break. But in order to put the glass back together—to decrease the glass's entropy—someone has to do work.

› Practice Problems

Multiple Choice:

1. To increase the diameter of an aluminum ring from 50.0 mm to 50.1 mm, the temperature of the ring must increase by 80°C. What temperature change would be necessary to increase the diameter of an aluminum ring from 100.0 mm to 100.1 mm?

   (A) 20°C
   (B) 40°C
   (C) 80°C
   (D) 110°C
   (E) 160°C

2. A gas is enclosed in a metal container with a moveable piston on top. Heat is added to the gas by placing a candle flame in contact with the container's bottom. Which of the following is true about the temperature of the gas?

   (A) The temperature must go up if the piston remains stationary.
   (B) The temperature must go up if the piston is pulled out dramatically.
   (C) The temperature must go up no matter what happens to the piston.
   (D) The temperature must go down no matter what happens to the piston.
   (E) The temperature must go down if the piston is compressed dramatically.

3. A small heat engine operates using a pan of 100°C boiling water as the high temperature reservoir and the atmosphere as a low temperature reservoir. Assuming ideal behavior, how much more efficient is the engine on a cold, 0°C day than on a warm, 20°C day?

   (A) 1.2 times as efficient
   (B) 2 times as efficient
   (C) 20 times as efficient
   (D) infinitely more efficient
   (E) just as efficient

4. A 1-m³ container contains 10 moles of ideal gas at room temperature. At what fraction of atmospheric pressure is the gas inside the container?

   (A) ¹⁄₄₀ atm
   (B) ¹⁄₂₀ atm
   (C) ¹⁄₁₀ atm
   (D) ¹⁄₄ atm
   (E) ¹⁄₂ atm

Free Response:

5. A small container of gas undergoes a thermodynamic cycle. The gas begins at room temperature. First, the gas expands isobarically until its volume has doubled. Second, the gas expands adiabatically. Third, the gas is cooled isobarically; finally, the gas is compressed adiabatically until it returns to its original state.

   (a) The initial state of the gas is indicated on the PV diagram below. Sketch this process on the graph.

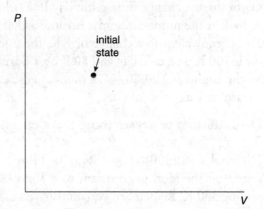

   (b) Is the temperature of the gas greater right before or right after the adiabatic expansion? Justify your answer.
   (c) Is heat added to or removed from the gas in one cycle?
   (d) Does this gas act as a refrigerator or a heat engine?

> Solutions to Practice Problems

1. B—Solve the formula for length expansion for temperature change:

$$\Delta T = \frac{\Delta L}{\alpha L_o}.$$

The only term on the right-hand side that is different from the original situation is L_0, which has doubled (from 50.0 mm to 100.0 mm). If you double the denominator of an equation, the entire equation is cut in half—thus the answer is 40°C.

2. A—Use the first law of thermodynamics, $\Delta U = Q + W$. The candle adds heat to the gas, so Q is positive. Internal energy is directly related to temperature, so if ΔU is positive, then temperature goes up (and vice versa). Here we can be sure that ΔU is positive if the work done on the gas is either positive or zero. The only possible answer is A—for B, the work done on the gas is negative because the gas expands. (Note that just because we add heat to a gas does NOT mean that temperature automatically goes up!)

3. A—The equation for efficiency of an ideal heat engine is

$$\frac{T_H - T_C}{T_H}.$$

The temperatures must be in kelvins. The denominator doesn't change from a hot day to a cold day, so look at the numerator only. Because a change of 1°C is equivalent to a change of 1 K, the numerator is 100 K on a cold day and 80 K on a warm day. So the engine is 100/80 ~ 1.2 times as efficient on a warm day as on a cold day.

4. D—Estimate the answer using the ideal gas law. Solving for pressure, $P = nRT/V$. Plug in: (10 moles)(8)(300)/(1) = 24,000 N/m². (Just round off the ideal gas constant to 8 J/mol·K and assume 300 K. Remember, you don't have a calculator, so make computation easy.) Well, atmospheric pressure is listed on the constant sheet and is 100,000 N/m². Thus, the pressure of the container is about $1/4$ of atmospheric.

5. (a)

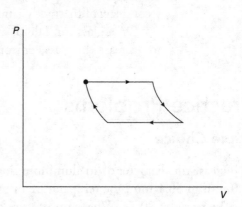

(b) The temperature is greater right before the expansion. By definition, in an adiabatic process, no heat is added or removed. But because the gas expanded, work was done *by* the gas, meaning the W term in the first law of thermodynamics is negative. So, by $\Delta U = Q + W$, $Q = 0$ and W is negative; so ΔU is negative as well. Internal energy is directly related to temperature. Therefore, because internal energy decreases, so does temperature.

(c) In a full cycle, the gas begins and ends at the same state. So the total change in internal energy is *zero*. Now consider the total work done on or by the gas. When the gas expands in the first and second process, there's more area under the graph than when the gas compresses in the second and third processes. So, the gas does more work expanding than compressing; the net work is thus done *by* the gas, and is *negative*. To get no change in internal energy, the Q term in the first law of thermodynamics must be positive; heat must be added to the gas.

(d) This is a heat engine. Heat is added to the gas, and net work is done by the gas; that's what a heat engine does. (In a refrigerator, net work is done on the gas, and heat is removed from the gas.)

❯ Rapid Review

- Heat is a form of energy that can be transferred from one body to another. The internal energy of an object is the sum of the kinetic energies of all the molecules that make up that object. An object's temperature is directly related to average kinetic energy per molecule of the object.

- Two bodies with the same temperature do not necessarily have the same internal energy, and vice versa.

- When objects are heated, they expand. The linear expansion of an object is proportional to the initial length of the object and the temperature by which it is heated.

- The pressure and volume of an ideal gas are related by the ideal gas law. The law can either be written as $PV = nRT$, or $PV = Nk_BT$.

- The first law of thermodynamics states that the change in a gas's internal energy equals the heat added to the gas plus the work done on the gas.

- PV diagrams illustrate the relationship between a gas's pressure and volume as it undergoes a process. Four types of processes a gas can undergo are (1) isothermal [temperature stays constant]; (2) adiabatic [no heat is transferred to or from the gas]; (3) isobaric [pressure stays constant]; and (4) isochoric [volume stays constant].

- The second law of thermodynamics says that heat flows naturally from a hot object to a cold object.

- A heat engine works by letting heat flow from hot stuff to cold stuff; some of the flowing heat is used to do work.

- The efficiency of a heat engine is always less than 1. The real efficiency of a heat engine equals the work done by the engine divided by the amount of input heat. The ideal efficiency of a heat engine depends only on the temperatures of the hot and cold reservoirs.

- A refrigerator is like a heat engine, except in a refrigerator, work is done on a system to make cold stuff hotter.

- Entropy is a measure of disorder. The second law of thermodynamics says that the entropy of a system cannot decrease unless work is done on that system.

CHAPTER > 18

Fluid Mechanics

IN THIS CHAPTER

Summary: The Physics B exam discusses both static and flowing fluids. The five principles listed under "key ideas" must be understood.

Key Ideas

✪ Three principles for static fluids:

(a) The pressure in a static column depends on surface pressure and the column's height.

(b) The buoyant force on an object is equal to the weight of the displaced fluid.

(c) Pascal's principle states that pressure changes in an enclosed fluid are the same throughout, leading to the utility of hydraulic lifts.

✪ Two principles for flowing fluids:

(a) The continuity equation is a statement of conservation of mass—what flows into a pipe must flow out if a fluid is incompressible.

(b) Bernoulli's equation is a statement of energy conservation.

Relevant Equations

Pressure in a static column:

$$P = p_0 + \rho g h$$

Buoyant force (Archimedes' principle):

$$F_B = \rho_{\text{fluid}} V_{\text{submerged}} g$$

Definition of pressure:

$$P = \frac{F}{A}$$

Fluid flow continuity:

$$A_1 v_1 = A_2 v_2$$

Bernoulli's equation:

$$P_1 + \rho g y_1 + \tfrac{1}{2}\rho v_1^2 = P_2 + \rho g y_2 + \tfrac{1}{2}\rho v_2^2$$

If you look back over the previous nine chapters, you might notice a striking omission: we've talked a lot about solid objects (like blocks and pulleys), and we've talked a bit about gases (when we discussed thermodynamics), but we haven't yet introduced liquids . . . until now.

Fluid mechanics is, in its simplest sense, the study of how liquids behave. You could read whole books about this topic, but fortunately, for the AP exam, you only need to learn five relatively simple concepts. Three of these concepts apply to static fluids, which are fluids that aren't moving; the other two concepts apply to flowing fluids, which are fluids that, clearly, are moving. We start with static fluids, and then we talk about flowing fluids toward the end of the chapter.

Pressure and Density

We first talked about **pressure** when we introduced gases in Chapter 18 (remember the ideal gas law?).

$$\boxed{\text{Pressure} = \frac{\text{Force}}{\text{Area}}}$$

A fluid, just like a gas, exerts a pressure outward in all directions on the sides of the container holding it. The formula in the box tells us that the pressure exerted by a fluid is equal to the amount of force with which it pushes on the walls of its container divided by the surface area of those walls. Recall that the standard units of pressure are newtons per square meter, which are also called pascals.

Besides pressure, the other important characteristic of a fluid is its **density**, abbreviated as the Greek letter ρ (rho).

$$\boxed{\rho = \frac{\text{Mass}}{\text{Volume}}}$$

As you could guess, the standard units of density are kg/m³. You probably learned about density in sixth grade science, so this should be a familiar topic for you. But make sure you've memorized the definition of density, because it's critical for studying fluids.

Specific Gravity

It is useful to know the density of water, just because water is such a common fluid. So let us ask you, "What is the density of water?"

You said, "1," right? And being the brilliant physics ace that you are, you quickly put some units on your answer: "1 g/cm³." But look back at our definition of density!

The standard units of density are kg/m³, so your answer would be more useful in these units. The best answer, then, is that water has a density of 1,000 kg/m³.

Because water is so ubiquitous, the density of a substance is often compared to water; for instance, alcohol is eight tenths as dense as water; lead is eleven times as dense. We call this comparison the **specific gravity** of a substance—lead's specific gravity is 11 because it is 11 times more dense than water. So its density in standard units is 11,000 kg/m³.

> **Exam tip from an AP Physics veteran:**
> Though the density of water will be given wherever it's necessary, it's still worth remembering as a point of comparison: 1,000 kg/m³. You know anything with a greater density will sink in water, anything with less density will float.
>
> —*Bret, high school senior*

Density and Mass

When dealing with fluids, we don't generally talk about mass—we prefer to talk about density and volume. But, of course, mass is related to these two measurements: *mass = density · volume.*

So the mass of any fluid equals ρV. And because weight equals *mg*, the weight of a fluid can be written as $\rho V g$.

Pressure in a Static Fluid Column

Consider a column of water in a container, such as the one shown in Figure 18.1.

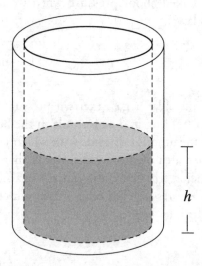

Figure 18.1   Column of water in a container.

The water has significant weight, all pushing down on the bottom of the container (as well as out on the sides). So we expect water pressure to increase as the water gets deeper. Sure enough, the pressure due to the column of fluid of density ρ is given by this formula:

$$P = P_0 + \rho g h$$

In this formula, P_0 is the pressure at the top of the column, and h is the vertical height of the column.

Note that h is a *vertical* height, and it does not depend on the width of the container, or even the shape of the container. A town's water tower is usually located at the highest available elevation, because the water pressure in someone's sink depends only on the altitude difference between the sink and the tower. The horizontal distance to the tower is irrelevant.

P_0 represents the pressure at the top of the column, which is often atmospheric pressure. But not always! For example, if you put your finger on top of a straw and lift up some water, the pressure at the *bottom* of the straw is atmospheric (because that part of the water is open to the atmosphere). So, the pressure at the top of the column P_0 must be *less* than atmospheric pressure by $P = P_0 + \rho gh$.

The pressure $P = P_0 + \rho gh$ is called the absolute pressure of a fluid. Gauge pressure refers to just the ρgh part of the absolute pressure. These are terms to know because they might show up on the AP exam.

The varsity football team's water cooler is a plastic cylinder 70 cm high with a 30-cm radius. A 2-cm diameter hole is cut near the bottom for water to flow out. This hole is plugged. What is the force on the plug due to the water (density = 1,000 kg/m^3) in the cooler when the cooler is full?

Let's start by calculating the water pressure at the bottom of the container: $P = P_0 + \rho gh$. P_0 is the pressure at the top of the container, which we'll assume is atmospheric pressure (atmospheric pressure is available on the constant sheet). And h is the depth of the cooler, 70 cm, which in standard units is 0.70 m.

$$P = (1 \times 10^5 \text{ N/m}^2) + (1,000 \text{ kg/m}^3)(10 \text{ N/kg})(0.70 \text{ m})$$

$$P = 107,000 \text{ N/m}^2$$

This pressure acts outward on the container everywhere at the bottom. This means the pressure pushes down on the container's bottom, and out on the container's sides near the bottom. So this pressure pushes on the plug.

How do we get the *force* on the plug? Since pressure is force/area, force is just pressure times area (and the area of the plug is πr^2, where r is the radius of the plug, which is 1 cm):

$$F = 107,000 \text{ N/m}^2 \cdot \pi (0.01 \text{ m})^2 = 34 \text{ N}$$

That's about 6 pounds of force—that plug had better be wedged in well! This is why you've usually got to push hard on the button at the bottom of a full water cooler to get anything out—you're opposing several pounds of force.

Buoyancy and Archimedes' Principle

When an object is submerged wholly or partially in a fluid, the object experiences an upward force called a buoyant force. This leads us to Archimedes' principle, which says, in words:

> The buoyant force on an object is equal to the weight of the fluid displaced by that object.

In mathematics, Archimedes' principle says:

$$F_B = (\rho_{\text{fluid}})(V_{\text{submerged}})\,g$$

Let's investigate this statement a bit. The weight of anything is *mg*. So the weight of the displaced fluid is *mg*, where *m* is the mass of this fluid. But in fluid mechanics, we like to write mass as ρV. Here, because we're talking about the mass of fluid, ρ refers to the density of the *fluid* (NOT the submerged mass). And because only the part of an object that is underwater displaces fluid, *V* represents only the *submerged* volume of an object.

It is extremely useful to use subscripts liberally when applying Archimedes' principle. Otherwise it's easy to get confused as to what volume, what density, and what mass you are dealing with.

So here's an example.

> We often hear the phrase, "just the tip of the iceberg." To investigate the origin of the phrase, calculate the fraction of an iceberg's volume that is visible above the water's surface. Ice has a specific gravity of 0.9.

Let's start by drawing a free-body diagram of the iceberg. The only forces acting on the iceberg are the buoyant force and the iceberg's weight.

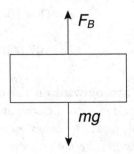

Since the iceberg is stationary, it is in equilibrium, so $F_B = mg$. We can use Archimedes' principle to rewrite the buoyant force as

$$F_B = (\rho_{\text{water}})(V_{\text{submerged}})g$$

And we know that the mass of the iceberg equals its density times its volume:

$$m_{\text{iceberg}} = (\rho_{\text{iceberg}})(V_{\text{iceberg}})$$

We can now rewrite the equation, $F_B = mg$:

$$(\rho_{\text{water}})(V_{\text{submerged}})g = (\rho_{\text{iceberg}})(V_{\text{iceberg}})g$$

To find the fraction of the iceberg submerged, we simply take the ratio of the submerged volume to the total volume.

$$\frac{V_{\text{submerged}}}{V_{\text{iceberg}}} = \frac{\rho_{\text{iceberg}}}{\rho_{\text{water}}}$$

The right-hand side works out to 0.9, because a specific gravity of 0.9 *means* that the iceberg's density over water's density is 0.9. Therefore, the left-hand side of the equation must equal 0.9, meaning that 90% of the iceberg is submerged. So 10% of the iceberg's volume is visible. Just the tip of the iceberg!

Pascal's Principle

When fluid is in a closed container, it exerts a pressure outward on every surface of its container. But what happens when an external force is applied on the container? This is where **Pascal's principle** comes in.

KEY IDEA

> If a force is applied somewhere on a container holding a fluid, the pressure increases *everywhere* in the fluid, not just where the force is applied.

This principle is most useful for hydraulic systems in machines. For example:

> Consider the hydraulic jack diagramed below. A person stands on a piston that pushes down on a thin cylinder full of water. This cylinder is connected via pipes to a wide platform on top of which rests a 1-ton (1000 kg) car. The area of the platform under the car is 25 m²; the person stands on a 0.3 m² piston. What is the lightest weight of a person who could successfully lift the car?
>
>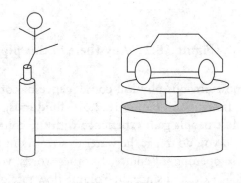

We have an enclosed fluid, so Pascal's principle insists that the pressure applied by the person should also be applied equally to the car. Well, we know what pressure will be necessary to lift the car: Pressure is force/area, or the weight of the car over the area of the car's platform:

$$P = 10{,}000 \text{ N}/25 \text{ m}^2 = 400 \text{ Pa (recall that a pascal, Pa, is a newton per m}^2\text{)}.$$

So the person needs to provide an equal pressure. The person's weight is applied over a much smaller area, only 0.3 m². So set 400 Pa equal to the person's weight divided by 0.3 m².

Flowing Fluids

So far we've only talked about stationary fluids. But what if a fluid is moving through a pipe? There are two questions that then arise: first, how do we find the velocity of the flow, and second, how do we find the fluid's pressure? To answer the first question, we'll turn to the continuity principle, and to answer the second, we'll use Bernoulli's equation.

Continuity

Consider water flowing through a pipe. Imagine that the water always flows such that the entire pipe is full. Now imagine that you're standing in the pipe[1]—how much water flows past you in a given time?

We call the volume of fluid that flows past a point every second the volume flow rate, measured in units of m^3/s. This rate depends on two characteristics: the velocity of the fluid's flow, v, and the cross-sectional area of the pipe, A. Clearly, the faster the flow, and the wider the pipe, the more fluid flows past a point every second. To be mathematically precise,

Volume flow rate = Av.

If the pipe is full, then any volume of fluid that enters the pipe must eject an equal volume of fluid from the other end. (Think about it—if this weren't the case, then the fluid would have to compress, or burst the pipe. Once either of these things happens, the principle of continuity is void.) So, by definition, *the volume flow rate is equal at all points within an isolated stream of fluid.* This statement is known as the principle of continuity. For positions "1" and "2" in Figure 18.2, for example, we can write $A_1 v_1 = A_2 v_2$.

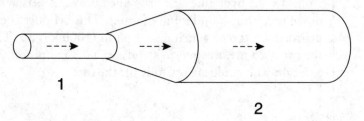

Figure 18.2 Anywhere in this pipe, the volume flow is the same.

The most obvious physical consequence of continuity is that where a pipe is narrow, the flow is faster. So in Figure 18.2, the fluid must move faster in the narrower section labeled "1." Most people gain experience with the continuity principle when using a garden hose. What do you do to get the water to stream out faster? You use your thumb to cover part of the hose opening, of course. In other words, you are decreasing the cross-sectional area of the pipe's exit, and since the volume flow rate of the water can't change, the velocity of the flow must increase to balance the decrease in the area.

Bernoulli's Equation

Bernoulli's equation is probably the longest equation you need to know for the AP exam, but fortunately, you don't need to know how to derive it. However, what you should know, conceptually, is that it's really just an application of conservation of energy.

$$\boxed{P_1 + \rho g y_1 + \tfrac{1}{2}\rho v_1^2 = P_2 + \rho g y_2 + \tfrac{1}{2}\rho v_2^2}$$

---

[1]Wearing a scuba mask, of course.

Bernoulli's equation is useful whenever you have a fluid flowing from point "1" to point "2." The flow can be fast or slow; it can be vertical or horizontal; it can be open to the atmosphere or enclosed in a pipe. (What a versatile equation!) P is the pressure of the fluid at the specified point, y is the vertical height at the specified point, ρ is the fluid's density, and v is the speed of the flow at the specified point.

Too many terms to memorize? Absolutely not . . . but a mnemonic device might still be helpful. So here's how we remember it. First, we're dealing with pressures, so obviously there should be a pressure term on each side of the equation. Second, we said that Bernoulli's equation is an application of energy conservation, so each side of the equation will contain a term that looks kind of like potential energy and a term that looks kind of like kinetic energy. (Note that the units of every term are N/m^2.)

Principal Consequence of Bernoulli's Equation

The most important result that can be derived from Bernoulli's equation is this: generally, where flow is faster, pressure is lower.

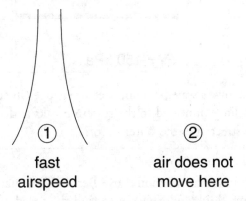

① fast airspeed ② air does not move here

Figure 18.3   Blowing air between two hanging sheets of paper.

Take two pieces of paper and hold them parallel to each other, a couple centimeters apart. Now get someone with strong lungs—a trombone player, perhaps—to blow a fast, focused stream of air between the sheets of paper. He will blow the sheets apart, right?

No! Try the experiment yourself! You'll see that the pieces of paper actually move together. To see why, let's look at the terms in Bernoulli's equation.[2] Take point "1" to be between the sheets of paper; take point "2" to be well away from the sheets of paper, out where the air is hardly moving, but at the same level as the sheets of paper so that y_1 and y_2 are the same. (See Figure 18.3.) We can cancel the y terms and solve for the pressure in between the sheets of paper:

$$P_1 = P_2 + (\tfrac{1}{2}\,\rho v_2^2 - \tfrac{1}{2}\,\rho v_1^2)$$

Far away from the pieces of paper, the air is barely moving (so v_2 is just about zero), and the pressure P_2 must be the ambient atmospheric pressure. But between the sheets, the air is moving quickly. So look at the terms in parentheses: v_1 is substantially bigger than v_2, so the part in parentheses gives a negative quantity. This means that to get P_1, we must *subtract* something from atmospheric pressure. The pressure between the pieces of paper is, therefore, *less than* atmospheric pressure, so the air pushes the pieces of paper together.

---

[2]Yes, Bernoulli's equation *can* be applied to gases as well as liquids, as long as the gas is obeying the continuity principle and not compressing. This is a good assumption in most cases you'll deal with on the AP exam.

This effect can be seen in numerous situations. A shower curtain is pushed in toward you when you have the water running past it; an airplane is pushed up when the air rushes faster over top of the wing than below the wing; a curveball curves because the ball's spin causes the air on one side to move faster than the air on the other side.

In general, solving problems with Bernoulli's equation is relatively straightforward. Yes, the equation has six terms: one with pressure, one with height, and one with speed for each of two positions. But in almost all cases you will be able to get rid of several of these terms, either because the term itself is equal to zero, or because it will cancel with an identical term on the other side of the equation. For example:

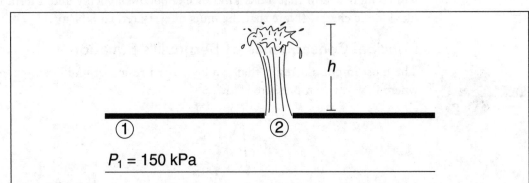

In a town's water system, pressure gauges in still water at street level read 150 kPa. If a pipeline connected to the system breaks and shoots water straight up, how high above the street does the water shoot?

We know that we should use Bernoulli's equation because we are given the pressure of a flowing fluid, and we want to find the speed of the flow. Then, kinematics can be used to find the height of the fountain.

Let point "1" be in still water where the 150 kPa reading was made, and point "2" will be where the pipe broke. Start by noting which terms in Bernoulli's equation will cancel: both points are at street level, so $y_1 = y_2$ and the height terms go away; the velocity at point "1" is zero because point "1" is in "still" water.

Now the pressure terms: If a *gauge* reads 150 kPa at point "1," this is obviously a reading of *gauge* pressure, which you recall means the pressure over and above atmospheric pressure. The pressure where the pipe broke must be atmospheric.

Exam tip from an AP Physics veteran:
Anytime a liquid is exposed to the atmosphere, its pressure must be atmospheric.
　　　　　　　　　　　—*Paul, former AP Physics student and current physics teacher*

Of the six terms in Bernoulli's equation, we have gotten rid of three:

$$P_1 + \rho g y_1 + \tfrac{1}{2}\rho v_1^2 = P_2 + \rho g y_2 + \tfrac{1}{2}\rho v_2^2.$$

This leaves

$$\tfrac{1}{2}\rho v_2^2 = P_1 - P_2.$$

P_1 is equal to $P_{atm} + 150$ kPa; P_2 is just atmospheric pressure:

$$\tfrac{1}{2}\rho v_2^2 = P_{atm} + 150\,\text{kPa} - P_{atm}.$$

The atmospheric pressure terms go away. Plug in the density of water (1000 kg/m³), change 150 kPa to 150,000 Pa, and solve for velocity:

$$v_2 = 17 \text{ m/s}.$$

Finally, we use kinematics to find the height of the water stream. Start by writing out the table of variables, then use the appropriate equation.

v_0	17 m/s
v_f	0 m/s
$x - x_0$	?
a	−10 m/s²
t	?

$$v_f^2 = v_0^2 + 2a(x - x_0)$$

$x - x_0 = 15$ m high, about the height of a four- or five-story building.

› Practice Problems

Multiple Choice:

1. A person sips a drink through a straw. At which of the following three positions is the pressure lowest?

   I. Inside the person's mouth
   II. At the surface of the drink
   III. At the bottom of the drink

   (A) only at position I
   (B) only at position II
   (C) only at position III
   (D) both at positions I and III
   (E) both at positions I and II

2. The circulatory system can be modeled as an inter-connected network of flexible pipes (the arteries and veins) through which a pump (the heart) causes blood to flow. Which of the following actions, while keeping all other aspects of the system the same, would NOT cause the velocity of the blood to increase inside a vein?

   (A) expanding the vein's diameter
   (B) cutting off blood flow to some other area of the body
   (C) increasing the heart rate
   (D) increasing the total amount of blood in the circulatory system
   (E) increasing the pressure difference between ends of the vein

3. A pirate ship hides out in a small inshore lake. It carries twenty ill-gotten treasure chests in its hold. But lo, on the horizon the lookout spies a gunboat. To get away, the pirate captain orders the heavy treasure chests jettisoned. The chests sink to the lake bottom. What happens to the water level of the lake?

   (A) The water level drops.
   (B) The water level rises.
   (C) The water level does not change.

4. Brian saves 2-liter soda bottles so that he can construct a raft and float out onto Haverford College's Duck Pond. If Brian has a mass of 80 kg, what minimum number of bottles is necessary to support him? The density of water is 1000 kg/m³, and 1000 L = 1 m³.

   (A) 1600 bottles
   (B) 800 bottles
   (C) 200 bottles
   (D) 40 bottles
   (E) 4 bottles

Free Response:

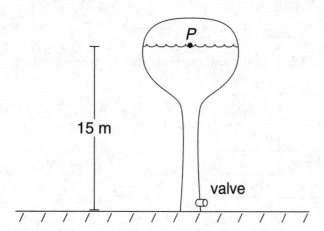

5. The water tower in the drawing above is drained by a pipe that extends to the ground. The amount of water in the top spherical portion of the tank is significantly greater than the amount of water in the supporting column.

Information you may need: Density of water = 1000 kg/m³.

(a) What is the absolute pressure at the position of the valve if the valve is *closed*, assuming that the top surface of the water at point P is at atmospheric pressure?

(b) Now the valve is opened; thus, the pressure at the valve is forced to be atmospheric pressure. What is the speed of the water past the valve?

(c) Assuming that the radius of the circular valve opening is 10 cm, find the volume flow rate out of the valve.

(d) Considering that virtually all of the water is originally contained in the top spherical portion of the tank, estimate the initial volume of water contained by the water tower. Explain your reasoning thoroughly.

(e) Estimate how long it would take to drain the tank completely using this single valve.

> Solutions to Practice Problems

1. A—The fluid is pushed into the mouth by the atmospheric pressure. Because the surface of a drink is open to the atmosphere, the surface is at atmospheric pressure, and the pressure in the mouth must be lower than atmospheric.

2. A—Flow rate (volume of flow per second) is the area of a pipe times the speed of flow. So if we increase the volume of flow (choices B, C, and D), we increase the speed. By Bernoulli's equation, choice E also increases the fluid speed. Expanding the vein increases the area of the pipe, so if flow rate is constant, then the velocity must *decrease*.

3. A—When the treasure is floating in the boat, it displaces an amount of water equal to its weight. When the treasure is on the lake bottom, it displaces much less water, because the lake bottom supports most of the weight that the buoyant force was previously supporting. Thus, the lake level drops.

4. D—Since Brian will be floating in equilibrium, his weight must be equal to the buoyant force on him. The buoyant force is $\rho_{water}V_{submerged}g$, and must equal Brian's weight of 800 N. Solving for $V_{submerged}$, we find he needs to displace 8/100 of a cubic meter. Converting to liters, he needs to displace 80 L of water, or 40 bottles. (We would suggest that he use, say, twice this many bottles—then the raft would float only half submerged, and he would stay drier.)

5. (a) If the valve is closed, we have a static column of water, and the problem reduces to one just like the example in the chapter. $P = P_0 + \rho gh = 10^5$ N/m² + (1000 kg/m³)(10 N/kg)(15 m) = 250,000 N/m² (atmospheric pressure is given on the constant sheet).

(b) Use Bernoulli's equation for a flowing fluid. Choose point P and the valve as our two positions. The pressure at both positions is now atmospheric, so the pressure terms go away. Choose the height of the valve to be zero. The speed of the water at the top is just about zero, too. So, four of the six terms in Bernoulli's equation cancel! The equation becomes $\rho gy_{top} = 1/2\,\rho v_{bottom}^2$. Solving, $v_{bottom} = 17$ m/s.

(c) Flow rate is defined as Av. The cross-sectional area of the pipe is $\pi(0.1 \text{ m})^2 = 0.031$ m². (Don't forget to use meters, not centimeters!) So the volume flow rate is 0.53 m³/s.

(d) The volume of the spherical portion of the tank can be estimated as $(4/3)\pi r^3$ (this equation is on the equation sheet), where the radius of the tank looks to be somewhere around 2 or 3 meters. Depending what actual radius you choose, this gives a volume of about 100 m³. The tank looks to be something like 3/4 full . . . So call it 70 m³. (Any correct reasoning that leads to a volume between, say, 30–300 m³ should be accepted.)

(e) The flow rate is 0.53 m³/s; we need to drain 70 m³. So, this will take 70/0.53 = 140 seconds, or about two minutes. [Again, your answer should be counted as correct if the reasoning is correct and the answer is consistent with part (d).]

› Rapid Review

- Pressure equals force divided by the area over which the force is applied.

- Density equals the mass of a substance divided by its volume. The ratio of a substance's density to the density of water is called the substance's "specific gravity."

- The absolute pressure exerted by a column of a fluid equals the pressure at the top of the column plus the gauge pressure—which equals $\rho g h$.

- The buoyant force on an object is equal to the weight of the fluid displaced by that object.

- If a force is applied somewhere on a container holding a fluid, the pressure increases everywhere in the fluid, not just where the force is applied.

- The volume flow rate—the volume of fluid flowing past a point per second—is the same everywhere in an isolated fluid stream. This means that when the diameter of a pipe decreases, the velocity of the fluid stream increases, and vice versa.

- Bernoulli's equation is a statement of conservation of energy. Among other things, it tells us that when the velocity of a fluid stream increases, the pressure decreases.

CHAPTER 19

Electrostatics

IN THIS CHAPTER

Summary: An electric field provides a force on a charged particle. Electric potential, also called voltage, provides energy to a charged particle. Once you know the force or energy experienced by a charged particle, Newtonian mechanics (i.e., kinematics, conservation of energy, etc.) can be applied to predict the particle's motion.

Key Ideas

✪ The electric force on a charged particle is qE, regardless of what produces the electric field. The electric potential energy of a charged particle is qV.

✪ Positive charges are forced in the direction of an electric field; negative charges, opposite the field.

✪ Positive charges are forced from high to low potential; negative charges, low to high.

✪ Point charges produce non-uniform electric fields. Parallel plates produce a uniform electric field between them.

✪ Electric field is a vector, and electric potential is a scalar.

Relevant Equations

Electric force on a charge in an electric field:

$$F = qE$$

Electric field produced by a point charge[1]:

$$E = \frac{kQ}{r^2}$$

Electric field produced by parallel plates:

$$E = \frac{V}{d}$$

Electric potential energy in terms of voltage:

$$PE = qV$$

Voltage produced by a point charge:

$$V = \frac{kQ}{r}$$

Charge stored on a capacitor:

$$Q = CV$$

Capacitance of a parallel plate capacitor:

$$C = \frac{\varepsilon_0 A}{d}$$

Electricity literally holds the world together. Sure, gravity is pretty important, too, but the primary reason that the molecules in your body stick together is because of electric forces. A world without electrostatics would be no world at all.

This chapter introduces a lot of the vocabulary needed to discuss electricity, and it focuses on how to deal with electric charges that aren't moving: hence the name, electro*statics*. We'll look at moving charges in the next chapter, when we discuss circuits.

Electric Charge

All matter is made up of three types of particles: protons, neutrons, and electrons. Protons have an intrinsic property called "positive charge." Neutrons don't contain any charge, and electrons have a property called "negative charge."

The unit of charge is the coulomb, abbreviated C. One proton has a charge of 1.6×10^{-19} coulombs.

Most objects that we encounter in our daily lives are electrically neutral—things like couches, for instance, or trees, or bison. These objects contain as many positive charges as negative charges. In other words, they contain as many protons as electrons.

When an object has more protons than electrons, though, it is described as "positively charged"; and when it has more electrons than protons, it is described as "negatively charged." The reason that big objects like couches and trees and bison don't behave like charged particles is because they contain so many bazillions of protons and electrons that an extra few here or there won't really make much of a difference. So even though they might have a slight electric charge, that charge would be much too small, relatively speaking, to detect.

Tiny objects, like atoms, more commonly carry a measurable electric charge, because they have so few protons and electrons that an extra electron, for example, would make a big difference. Of course, you can have very large charged objects. When you walk across a

---

[1]The actual equation sheet for the AP exam includes this equation: $F = \dfrac{kq_1 q_2}{r^2}$. You can see that this equation for the force between two point charges is an amalgamation of the first two equations above: Combine $F = qE$ with $E = \dfrac{k\dfrac{1}{4\pi\varepsilon_0}}{r^2}$ and you get $F = \dfrac{kq_1 q_2}{r^2}$.

carpeted floor in the winter, you pick up lots of extra charges and become a charged object yourself . . . until you touch a doorknob, at which point all the excess charge in your body travels through your finger and into the doorknob, causing you to feel a mild electric shock.

Electric charges follow a simple rule: *Like charges repel; opposite charges attract.* Two positively charged particles will try to get as far away from each other as possible, while a positively charged particle and a negatively charged particle will try to get as close as possible.

You can also have something called "induced charge." An induced charge occurs when an electrically neutral object becomes polarized—when negative charges pile up in one part of the object and positive charges pile up in another part of the object. The drawing in Figure 19.1 illustrates how you can create an induced charge in an object.

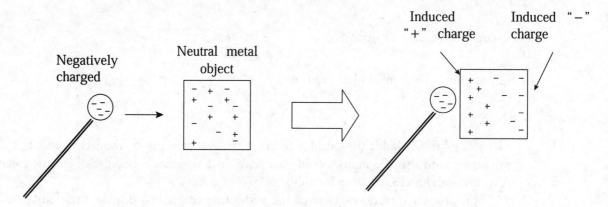

Figure 19.1   Creation of an induced charge.

Electric Fields

Before we talk about electric fields, we'll first define what a field, in general, is.

> **Field:** A property of a region of space that can apply a force to objects found in that region of space

A gravitational field is a property of the space that surrounds any massive object. There is a gravitational field that you are creating and which surrounds you, and this field extends infinitely into space. It is a weak field, though, which means that it doesn't affect other objects very much—you'd be surprised if everyday objects started flying toward each other because of gravitational attraction. The Earth, on the other hand, creates a strong gravitational field. Objects are continually being pulled toward the Earth's surface due to gravitational attraction. However, the farther you get from the center of the Earth, the weaker the gravitational field, and, correspondingly, the weaker the gravitational attraction you would feel.

An **electric field** is a bit more specific than a gravitational field: it only affects charged particles.

> **Electric Field:** A property of a region of space that applies a force to *charged* objects in that region of space. A charged particle in an electric field will experience an electric force.

Unlike a gravitational field, an electric field can either push or pull a charged particle, depending on the charge of the particle. Electric field is a vector; so, electric fields are always drawn as arrows.

Every point in an electric field has a certain value called, surprisingly enough, the "electric field value," or *E*, and this value tells you how strongly the electric field at that point would affect a charge. The units of *E* are newtons/coulomb, abbreviated N/C.

Force of an Electric Field

The force felt by a charged particle in an electric field is described by a simple equation:

$$F = qE$$

In other words, the force felt by a charged particle in an electric field is equal to the charge of the particle, q, multiplied by the electric field value, E.

> The direction of the force on a positive charge is in the same direction as the electric field; the direction of the force on a negative charge is opposite the electric field.

Let's try this equation on for size. Here's a sample problem:

An electron, a proton, and a neutron are each placed in a uniform electric field of magnitude 60 N/C, directed to the right. What is the magnitude and direction of the force exerted on each particle?

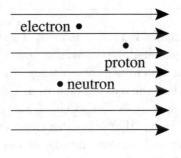

The solution here is nothing more than plug-and-chug into $F = qE$. Notice that we're dealing with a *uniform* electric field—the field lines are evenly spaced throughout the whole region. This means that, no matter where a particle is within the electric field, it always experiences an electric field of exactly 60 N/C.

Also note our problem-solving technique. To find the magnitude of the force, we plug in *just the magnitude* of the charge and the electric field—no negative signs allowed! To find the direction of the force, use the reasoning in the box above (positive charges are forced in the direction of the E field, negative charges opposite the E field).

Let's start with the electron, which has a charge of 1.6×10^{-19} C (no need to memorize, you can look this up on the constant sheet):

$$F = qE.$$
$$F = (1.6 \times 10^{-19} \text{ C})(60 \text{ N/C}).$$
$$F = 9.6 \times 10^{-18} \text{ N to the LEFT}$$

Now the proton:

$$F = (1.6 \times 10^{-19} \text{ C})(60 \text{ N/C}).$$
$$F = 9.6 \times 10^{-18} \text{ N to the RIGHT}$$

And finally the neutron:

$$F = (0 \text{ C})(60 \text{ N/C}) = 0 \text{ N}$$

Notice that the proton feels a force in the direction of the electric field, but the electron feels the same force in the opposite direction.

Don't state a force with a negative sign. Signs just indicate the direction of a force, anyway. So, just plug in the values for q and E, then state the direction of the force in words.

Electric Potential

When you hold an object up over your head, that object has gravitational potential energy. If you were to let it go, it would fall to the ground.

Similarly, a charged particle in an electric field can have electrical potential energy. For example, if you held a proton in your right hand and an electron in your left hand, those two particles would want to get to each other. Keeping them apart is like holding that object over your head; once you let the particles go, they'll travel toward each other just like the object would fall to the ground.

In addition to talking about electrical potential energy, we also talk about a concept called electric potential.

> **Electric Potential:** Potential energy provided by an electric field per unit charge; also called **voltage**

Electric potential is a scalar quantity. The units of electric potential are volts. 1 volt = 1 J/C.

Just as we use the term "zero of potential" in talking about gravitational potential, we can also use that term to talk about voltage. We cannot solve a problem that involves voltage unless we know where the zero of potential is. Often, the zero of electric potential is called "ground."

Unless it is otherwise specified, the zero of electric potential is assumed to be far, far away. This means that if you have two charged particles and you move them farther and farther from each another, ultimately, once they're infinitely far away from each other, they won't be able to feel each other's presence.

The electrical potential energy of a charged particle is given by this equation:

$$\text{PE} = qV$$

Here, q is the charge on the particle, and V is the voltage.

It is extremely important to note that electric potential and electric field are not the same thing. This example should clear things up:

> Three points, labeled A, B, and C, are found in a uniform electric field. At which point will a positron [a positively charged version of an electron] have the greatest electrical potential energy?
>
>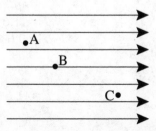

Electric field lines point in the direction that a positive charge will be forced, which means that our positron, when placed in this field, will be pushed from left to right. So, just as an object in Earth's gravitational field has greater potential energy when it is higher off the ground (think "*mgh*"), our positron will have the greatest electrical potential energy when it is farthest from where it wants to get to. The answer is A.

We hope you noticed that, even though the electric field was the same at all three points, the electric potential was different at each point.

How about another example?

A positron is given an initial velocity of 6×10^6 m/s to the right. It travels into a uniform electric field, directed to the left. As the positron enters the field, its electric potential is zero. What will be the electric potential at the point where the positron has a speed of 1×10^6 m/s?

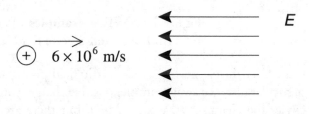

This is a rather simple conservation of energy problem, but it's dressed up to look like a really complicated electricity problem.

As with all conservation of energy problems, we'll start by writing our statement of conservation of energy.

$$KE_i + PE_i = KE_f + PE_f$$

Next, we'll fill in each term with the appropriate equations. Here the potential energy is not due to gravity (*mgh*), nor due to a spring ($1/2\ kx^2$). The potential energy is electric, so should be written as qV.

$$\tfrac{1}{2}\ mv_i^2 + qV_i = \tfrac{1}{2}\ mv_f^2 + qV_f$$

Finally, we'll plug in the corresponding values. The mass of a positron is exactly the same as the mass of an electron, and the charge of a positron has the same magnitude as the charge of an electron, except a positron's charge is positive. Both the mass and the charge of an electron are given to you on the "constants sheet." Also, the problem told us that the positron's initial potential V_i was zero.

$$\tfrac{1}{2}\ (9.1 \times 10^{-31}\ \text{kg})(6 \times 10^6\ \text{m/s})^2 + (1.6 \times 10^{-19}\ \text{C})(0) =$$
$$\tfrac{1}{2}\ (9.1 \times 10^{-31}\ \text{kg})(1 \times 10^6\ \text{m/s})^2 + (1.6 \times 10^{-19}\ \text{C})(V_f)$$

Solving for V_f, we find that V_f is about 100 V.

For *forces*, a negative sign simply indicates direction. For potentials, though, a negative sign is important. −300 V is less than −200 V, so a proton will seek out a −300 V position in preference to a −200 V position. So, be careful to use proper + and − signs when dealing with potential.

Just as you can draw electric field lines, you can also draw equipotential lines.

> **Equipotential Lines:** Lines that illustrate every point at which a charged particle would experience a given potential

Figure 19.2 shows a few examples of equipotential lines (shown with solid lines) and their relationship to electric field lines (shown with dotted lines):

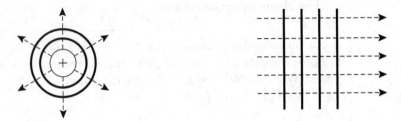

Figure 19.2   Two examples of equipotential lines (in bold) and electic field lines (dotted).

On the left in Figure 19.2, the electric field points away from the positive charge. At any particular distance away from the positive charge, you would find an equipotential line that circles the charge—we've drawn two, but there are an infinite number of equipotential lines around the charge. If the potential of the outermost equipotential line that we drew was, say, 10 V, then a charged particle placed anywhere on that equipotential line would experience a potential of 10 V.

On the right in Figure 19.2, we have a uniform electric field. Notice how the equipotential lines are drawn perpendicular to the electric field lines. In fact, equipotential lines are always drawn perpendicular to electric field lines, but when the field lines aren't parallel (as in the drawing on the left), this fact is harder to see.

Moving a charge from one equipotential line to another takes energy. Just imagine that you had an electron and you placed it on the innermost equipotential line in the drawing on the left. If you then wanted to move it to the outer equipotential line, you'd have to push pretty hard, because your electron would be trying to move toward, and not away from, the positive charge in the middle.

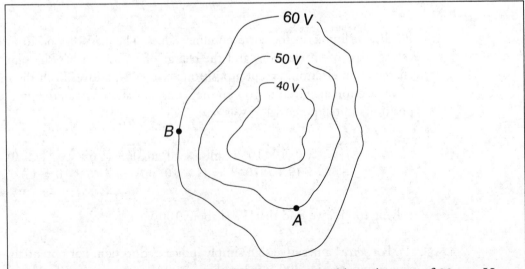

In the diagram above, point *A* and point *B* are separated by a distance of 30 cm. How much work must be done by an external force to move a proton from point *A* to point *B*?

The potential at point B is higher than at point A; so moving the positively charged proton from A to B requires work to change the proton's potential energy. The question here really is asking how much more potential energy the proton has at point B.

Well, potential energy is equal to qV; here, q is 1.6×10^{-19} C, the charge of a proton. The potential energy at point A is $(1.6 \times 10^{-19}$ C$)(50$ V$) = 8.0 \times 10^{-18}$ J; the potential energy at point B is $(1.6 \times 10^{-19}$ C$)(60$ V$) = 9.6 \times 10^{-18}$ J. Thus, the proton's potential is 1.6×10^{-18} J higher at point B, so it takes 1.6×10^{-18} J of work to move the proton there.

Um, didn't the problem say that points A and B were 30 cm apart? Yes, but that's irrelevant. Since we can see the equipotential lines, we know the potential energy of the proton at each point; the distance separating the lines is irrelevant.

Special Geometries for Electrostatics

There are two situations involving electric fields that are particularly nice because they can be described with some relatively easy formulas. Let's take a look:

Parallel Plates

If you take two metal plates, charge one positive and one negative, and then put them parallel to each other, you create a uniform electric field in the middle, as shown in Figure 19.3:

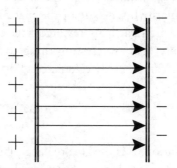

Figure 19.3   Electric field between charged, parallel plates.

The electric field between the plates has a magnitude of

$$E = \frac{V}{d}$$

V is the voltage difference between the plates, and d is the distance between the plates. Remember, this equation only works for parallel plates.

Charged parallel plates can be used to make a **capacitor**, which is a charge-storage device. When a capacitor is made from charged parallel plates, it is called, logically enough, a "parallel-plate capacitor." A schematic of this type of capacitor is shown in Figure 20.4.

The battery in Figure 19.4 provides a voltage across the plates; once you've charged the capacitor, you disconnect the battery. The space between the plates prevents any charges from jumping from one plate to the other while the capacitor is charged. When you want to discharge the capacitor, you just connect the two plates with a wire.

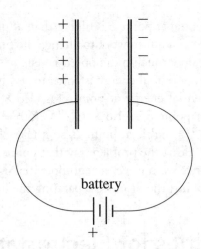

Figure 19.4   Basic parallel-plate capacitor.

The amount of charge that each plate can hold is described by the following equation:

$$Q = CV$$

Q is the charge on each plate, C is called the "capacitance," and V is the voltage across the plates. The capacitance is a property of the capacitor you are working with, and it is determined primarily by the size of the plates and the distance between the plates. The units of capacitance are farads, abbreviated F; 1 coulomb/volt = 1 farad.

The only really interesting thing to know about parallel-plate capacitors is that their capacitance can be easily calculated. The equation is:

$$C = \frac{\varepsilon_0 A}{d}$$

In this equation, A is the area of each plate (in m³), and d is the distance between the plates (in m). The term ε_0 (pronounced "epsilon-naught") is called the "permittivity of free space." The value of ε_0 is 8.84×10^{-12} C/V·m, which is listed on the constants sheet.

Capacitors become important when we work with circuits. So we'll see them again in Chapter 20.

Point Charges

As much as the writers of the AP exam like parallel plates, they *love* point charges. So you'll probably be using these next equations quite a lot on the test.

But, please don't go nuts . . . The formulas for force on a charge in an electric field ($F = qE$) and a charge's electrical potential energy ($PE = qV$) are your first recourse, your fundamental tools of electrostatics. On the AP exam, most electric fields are NOT produced by point charges! Only use the equations in this section when you have convinced yourself that a point charge is *creating* the electric field or the voltage in question.

First, the value of the electric field at some distance away from a point charge:

$$E = \frac{kQ}{r^2}$$

Q is the charge of your point charge, k is called the Coulomb's law constant ($k = 9 \times 10^9$ N·m²/C²), and r is the distance away from the point charge. *The field produced by a positive charge points away from the charge; the field produced by a negative charge points*

toward the charge. When finding an electric field with this equation, do NOT plug in the sign of the charge or use negative signs at all.

Second, the electric potential at some distance away from a point charge:

$$V = \frac{kQ}{r}$$

When using this equation, you *must* include a + or − sign on the charge creating the potential. (See Figure 19.5.)

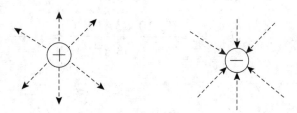

Figure 19.5   Electric field produced by point charges.

And third, the force that one point charge exerts on another point charge:

$$F = \frac{kQ_1Q_2}{r^2}$$

In this equation, Q_1 is the charge of one of the point charges, and Q_2 is the charge on the other one. This equation is known as Coulomb's law.

To get comfortable with these three equations, we'll provide you with a rather comprehensive problem.

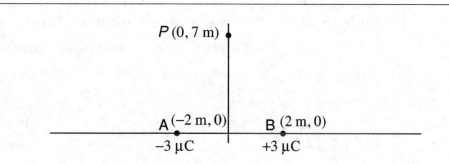

Two point charges, labeled "A" and "B", are located on the *x*-axis. "A" has a charge of −3 μC, and "B" has a charge of +3 μC. Initially, there is no charge at point *P*, which is located on the *y*-axis as shown in the diagram.

(1)   What is the electric field at point *P* due to charges "A" and "B"?
(2)   If an electron were placed at point *P*, what would be the magnitude and direction of the force exerted on the electron?
(3)   What is the electric potential at point *P* due to charges "A" and "B"?

Yikes! This is a monster problem. But if we take it one part at a time, you'll see that it's really not too bad.

Part 1—Electric Field

Electric field is a vector quantity. So we'll first find the electric field at point P due to charge "A," then we'll find the electric field due to charge "B," and then we'll add these two vector quantities. One note before we get started: to find r, the distance between points P and "A" or between P and "B," we'll have to use the Pythagorean theorem. We won't show you our work for that calculation, but you should if you were solving this on the AP exam.

$$E_{\text{due to "A"}} = \frac{(9\times10^9)(3\times10^{-6}\,\text{C})}{(\sqrt{53}\,\text{m})^2} = 510\,\frac{\text{N}}{\text{C}}, \text{ pointing } toward \text{ charge A}$$

$$E_{\text{due to "B"}} = \frac{(9\times10^9)(3\times10^{-6}\,\text{C})}{(\sqrt{53}\,\text{m})^2} = 510\,\frac{\text{N}}{\text{C}}, \text{ pointing } away\ from \text{ charge B}$$

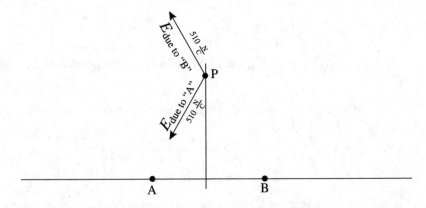

Note that we didn't plug in any negative signs! Rather, we calculated the magnitude of the electric field produced by each charge, and showed the direction on the diagram.

Now, to find the net electric field at point P, we must add the electric field vectors. This is made considerably simpler by the recognition that the y-components of the electric fields cancel . . . both of these vectors are pointed at the same angle, and both have the same magnitude. So, let's find just the x-component of one of the electric field vectors:

$$E_x = E\cos\theta, \text{where } \theta \text{ is measured from the horizontal.}$$

$E = 510$ N/C

$E_x\ (= 140 \text{ N/C})$

θ

Some quick trigonometry will find $\cos\theta$ . . . since $\cos\theta$ is defined as $\dfrac{\text{adjacent}}{\text{hypotenuse}}$, inspection of the diagram shows that $\cos\theta = \dfrac{2}{\sqrt{53}}$. So, the horizontal electric field $E_x = (510\text{ m})$ $\left(\dfrac{2}{\sqrt{53}}\right)$ . . . this gives 140 N/C.

And now finally, there are *TWO* of these horizontal electric fields adding together to the left—one due to charge "A" and one due to charge "B". The total electric field at point P, then, is

280 N/C, to the left

Part 2—Force

The work that we put into Part 1 makes this part easy. Once we have an electric field, it doesn't matter what caused the E field—just use the basic equation $F = qE$ to solve for the force on the electron, where q is the charge of the electron. So,

$$F = (1.6 \times 10^{-19} \text{ C}) \, 280 \text{ N/C} = 4.5 \times 10^{-17} \text{ N}$$

The direction of this force must be OPPOSITE the E field because the electron carries a negative charge; so, **to the right**.

Part 3—Potential

The nice thing about electric potential is that it is a scalar quantity, so we don't have to concern ourselves with vector components and other such headaches.

$$V_{\text{due to "A"}} = \frac{(9 \times 10^9)(-3 \times 10^{-6} \text{C})}{\sqrt{53} \text{ m}} = -3700 \text{ V}$$

$$V_{\text{due to "B"}} = \frac{(9 \times 10^9)(+3 \times 10^{-6} \text{C})}{\sqrt{53} \text{ m}} = +3700 \text{ V}$$

The potential at point P is just the sum of these two quantities. $V =$ zero!

Notice that when finding the electric potential due to point charges, you must include negative signs . . . negative potentials can cancel out positive potentials, as in this example.

› Practice Problems

Multiple Choice:

Questions 1 and 2

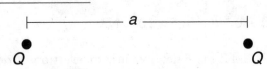

Two identical positive charges Q are separated by a distance a, as shown above.

1. What is the electric field at a point halfway between the two charges?

    (A) kQ/a^2
    (B) $2kQ/a^2$
    (C) zero
    (D) kQQ/a^2
    (E) $2kQ/a^2$

2. What is the electric potential at a point halfway between the two charges?

    (A) kQ/a
    (B) $2kQ/a$
    (C) zero
    (D) $4kQ/a$
    (E) $8kQ/a$

Questions 3 and 4

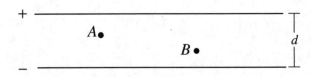

The diagram above shows two parallel metal plates that are separated by distance d. The potential difference between the plates is V. Point A is twice as far from the negative plate as is point B.

3. Which of the following statements about the electric potential between the plates is correct?

    (A) The electric potential is the same at points A and B.
    (B) The electric potential is two times larger at A than at B.
    (C) The electric potential is two times larger at B than at A.
    (D) The electric potential is four times larger at A than at B.
    (E) The electric potential is four times larger at B than at A.

4. Which of the following statements about the electric field between the plates is correct?

    (A) The electric field is the same at points A and B.
    (B) The electric field is two times larger at A than at B.
    (C) The electric field is two times larger at B than at A.
    (D) The electric field is four times larger at A than at B.
    (E) The electric field is four times larger at B than at A.

Free Response:

5. Two conducting metal spheres of different radii, as shown above, each have charge $-Q$.

    (a) Consider one of the spheres. Is the charge on that sphere likely to clump together or to spread out? Explain briefly.
    (b) Is charge more likely to stay inside the metal spheres or on the surface of the metal spheres? Explain briefly.
    (c) If the two spheres are connected by a metal wire, will charge flow from the big sphere to the little sphere, or from the little sphere to the big sphere? Explain briefly.
    (d) Which of the following two statements is correct? Explain briefly.
        i. If the two spheres are connected by a metal wire, charge will stop flowing when the electric field at the surface of each sphere is the same.
        ii. If the two spheres are connected by a metal wire, charge will stop flowing when the electric potential at the surface of each sphere is the same.
    (e) Explain how the correct statement you chose from part (d) is consistent with your answer to (c).

› Solutions to Practice Problems

1. **C**—Electric field is a *vector*. Look at the field at the center due to each charge. The field due to the left-hand charge points away from the positive charge; i.e., to the right; the field due to the right-hand charge points to the left. Because the charges are equal and are the same distance from the center point, the fields due to each charge have equal magnitudes. So the electric field vectors cancel! $E = 0$.

2. **D**—Electric potential is a *scalar*. Look at the potential at the center due to each charge: Each charge is distance $a/2$ from the center point, so the potential due to each is $kQ/(a/2)$, which works out to $2kQ/a$. The potentials due to both charges are positive, so add these potentials to get $4kQ/a$.

3. **B**—If the potential difference between plates is, say, 100 V, then we could say that one plate is at +100 V and the other is at zero V. So, the potential *must* change at points in between the plates. The electric field is uniform and equal to V/d (d is the distance between plates). Thus, the potential increases linearly between the plates, and A must have twice the potential as B.

4. **A**—The electric field by definition is *uniform* between parallel plates. This means the field must be the same everywhere inside the plates.

5. (a) Like charges repel, so the charges are more likely to spread out from each other as far as possible.

   (b) "Conducting spheres" mean that the charges are free to move anywhere within or onto the surface of the spheres. But because the charges try to get as far away from each other as possible, the charge will end up on the surface of the spheres. This is actually a property of conductors—charge will always reside on the surface of the conductor, not inside.

   (c) Charge will flow from the smaller sphere to the larger sphere. Following the reasoning from parts (a) and (b), the charges try to get as far away from each other as possible. Because both spheres initially carry the same charge, the charge is more concentrated on the smaller sphere; so the charge will flow to the bigger sphere to spread out. (The explanation that negative charge flows from low to high potential, and that potential is less negative at the surface of the bigger sphere, is also acceptable here.)

   (d) The charge will flow until the potential is equal on each sphere. By definition, negative charges flow from low to high potential. So, if the potentials of the spheres are equal, no more charge will flow.

   (e) The potential at the surface of each sphere is $-kQ/r$, where r is the radius of the sphere. Thus, the potential at the surface of the smaller sphere is initially more negative, and the charge will initially flow low-to-high potential to the larger sphere.

❯ Rapid Review

- Matter is made of protons, neutrons, and electrons. Protons are positively charged, neutrons have no charge, and electrons are negatively charged.

- Like charges repel, opposite charges attract.

- An induced charge can be created in an electrically neutral object by placing that object in an electric field.

- Electric field lines are drawn from positive charges toward negative charges. Where an electric field is stronger, the field lines are drawn closer together.

- The electric force on an object depends on both the object's charge and the electric field it is in.

- Unless stated otherwise, the zero of electric potential is at infinity.

- Equipotential lines show all the points where a charged object would feel the same electric force. They are always drawn perpendicular to electric field lines.

- The electric field between two charged parallel plates is constant. The electric field around a charged particle depends on the distance from the particle.

CHAPTER 20

Circuits

IN THIS CHAPTER

Summary: Electric charge flowing through a wire is called current. An electrical circuit is built to control current. In this chapter, you will learn how to predict the effects of current flow.

Key Ideas

✪ The current through series resistors is the same through each, whereas the voltage across series resistors adds to the total voltage.

✪ The voltage across parallel resistors is the same across each, whereas the current through parallel resistors adds to the total current.

✪ The brightness of a light bulb depends on the power dissipated by the bulb.

✪ A capacitor in a circuit blocks current and stores charge.

Relevant Equations

Definition of current:

$$I = \frac{\Delta Q}{\Delta t}$$

Resistance of a wire in terms of its properties:

$$R = \rho \frac{L}{A}$$

Ohm's law:

$$V = IR$$

Power in a circuit:

$$P = IV$$

In the last chapter, we talked about situations where electric charges don't move around very much. Isolated point charges, for example, just sit there creating an electric field. But what happens when you get a lot of charges all moving together? That, at its essence, is what goes on in a circuit.

Besides discussing circuits in general, this chapter presents a powerful problem-solving technique: the *V-I-R* chart. As with the chart of variables we used when solving kinematics problems, the *V-I-R* chart is an incredibly effective way to organize a problem that involves circuits. We hope you'll find it helpful.

Current

A circuit is simply any path that will allow charge to flow.

> **Current:** The flow of positive electric charge. In a circuit, the current is the amount of charge passing a given point per unit time.

A current is defined as the flow of positive charge. We don't think this makes sense, because electrons—and not protons or positrons—are what flow in a circuit. But physicists have their rationale, and no matter how wacky, we won't argue with it—the AP exam always uses this definition.

In more mathematical terms, current is defined as follows:

$$I = \frac{\Delta Q}{\Delta t}$$

What this means is that the current, *I*, equals the amount of charge flowing past a certain point divided by the time interval during which you're making your measurement. This definition tells us that current is measured in coulombs/second. 1 C/s = 1 ampere, abbreviated as 1 A.

Resistance and Ohm's Law

You've probably noticed that just about every circuit drawn in your physics book contains a battery. The reason most circuits contain a battery is because batteries create a potential difference between one end of the circuit and the other. In other words, if you connect the terminals of a battery with a wire, the part of the wire attached to the "+" terminal will have a higher electric potential than the part of the wire attached to the "−" terminal. And positive charge flows from high potential to low potential. So, in order to create a current, you need a battery. (See Figure 20.1.)

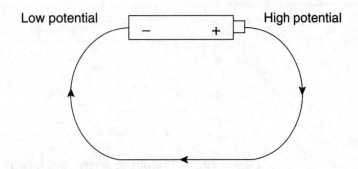

Figure 20.1 Flow of charge in a wire connected to a battery.

In general, the greater the potential difference between the terminals of the battery, the more current flows.

The amount of current that flows in a circuit is also determined by the resistance of the circuit.

> **Resistance:** A property of a circuit that resists the flow of current

Resistance is measured in ohms. 1 ohm is abbreviated as 1 Ω.

If we have some length of wire, then the resistance of that wire can be calculated. Three physical properties of the wire affect its resistance:

- The material the wire is made out of: the **resistivity**, ρ, of a material is an intrinsic property of that material. Good conducting materials, like gold, have low resistivities.[1]
- The length of the wire, L: the longer the wire, the more resistance it has.
- The cross-sectional area A of the wire: the wider the wire, the less resistance it has.

We put all of these properties together in the equation for resistance of a wire:

$$R = \rho \frac{L}{A}$$

Now, this equation is useful only when you need to calculate the resistance of a wire from scratch. Usually, on the AP exam or in the laboratory, you will be using resistors that have a pre-measured resistance.

> **Resistor:** Something you put in a circuit to change the circuit's resistance

Resistors are typically ceramic, a material that doesn't allow current to flow through it very easily. Another common type of resistor is the filament in a light bulb. When current flows into a light bulb, it gets held up in the filament. While it's hanging out in the filament, it makes the filament extremely hot, and the filament gives off light.

To understand resistance, an analogy is helpful. A circuit is like a network of pipes. The current is like the water that flows through the pipes, and the battery is like the pump that keeps the water flowing. If you wanted to impede the flow, you would add some narrow sections to your network of pipes. These narrow sections are your resistors. (See Figure 20.2.)

---

[1] Resistivity would be given on the AP exam if you need a value. Nothing here to memorize.

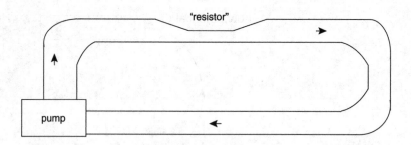

Figure 20.2 Plumbing system analogous to an electric circuit.

The way that a resistor (or a bunch of resistors) affects the current in a circuit is described by Ohm's law.

$$\boxed{\textbf{Ohm's law: } V = IR}$$

V is the voltage across the part of the circuit you're looking at, I is the current flowing through that part of the circuit, and R is the resistance in that part of the circuit. Ohm's law is the most important equation when it comes to circuits, so make sure you know it well.

When current flows through a resistor, electrical energy is being converted into heat energy. The rate at which this conversion occurs is called the power dissipated by a resistor. This power can be found with the equation

$$\boxed{P = IV}$$

This equation says that the power, P, dissipated in part of a circuit equals the current flowing through that part of the circuit multiplied by the voltage across that part of the circuit.

Using Ohm's law, it can be easily shown that $IV = I^2R = V^2/R$. It's only worth memorizing the first form of the equation, but any one of these could be useful.

Resistors in Series and in Parallel

In a circuit, resistors can either be arranged in series with one another or parallel to one another. Before we take a look at each type of arrangement, though, we need first to familiarize ourselves with circuit symbols, shown in Figure 20.3.

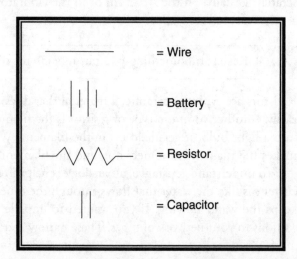

Figure 20.3 Common circuit symbols.

First, let's examine resistors in series. In this case, all the resistors are connected in a line, one after the other after the other:

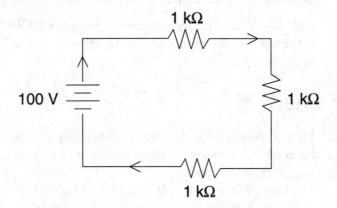

Figure 20.4 Example of series resistors.

To find the equivalent resistance of series resistors, we just add up all the individual resistors.

$$R_{eq} = R_1 + R_2 + R_3 \ldots$$

For the circuit in Figure 20.4, $R_{eq} = 3000\ \Omega$. In other words, using three 1000 Ω resistors in series produces the same total resistance as using one 3000 Ω resistor.

Parallel resistors are connected in such a way that you create several paths through which current can flow. For the resistors to be truly in parallel, the current must split, then immediately come back together.

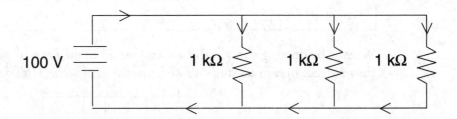

Figure 20.5 Example of parallel resistors.

The equivalent resistance of parallel resistors is found by this formula:

$$\frac{1}{R_{eq}} = \frac{1}{R_1} + \frac{1}{R_2} + \frac{1}{R_3} \ldots$$

For the circuit in Figure 20.5, the equivalent resistance is 333 Ω. So hooking up three 1000 Ω resistors in parallel produces the same total resistance as using one 333 Ω resistor. (Note that the equivalent resistance of parallel resistors is *less than* any individual resistor in the parallel combination.)

A Couple of Important Rules

Rule #1—When two resistors are connected in SERIES, the amount of current that flows through one resistor equals the amount of current that flows through the other resistor and is equal to the total current through both resistors.

Rule #2—When two resistors are connected in PARALLEL, the voltage across one resistor is the same as the voltage across the other resistor and is equal to the total voltage across both resistors.

The *V-I-R* Chart

Here it is—the trick that will make solving circuits a breeze. Use this method on your homework. Use this method on your quizzes and tests. But most of all, use this method on the AP exam. It works.

The easiest way to understand the *V-I-R* chart is to see it in action, so we'll go through a problem together, filling in the chart at each step along the way.

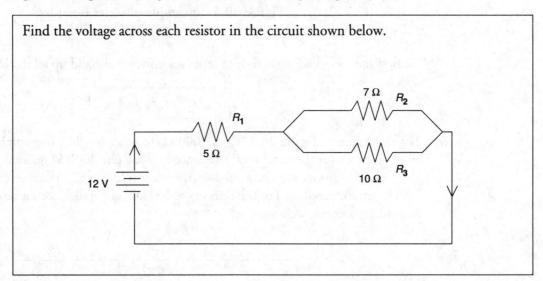

Find the voltage across each resistor in the circuit shown below.

We start by drawing our *V-I-R* chart, and we fill in the known values. Right now, we know the resistance of each resistor, and we know the total voltage (it's written next to the battery).

	V	*I*	*R*
R_1			$5\,\Omega$
R_2			$7\,\Omega$
R_3			$10\,\Omega$
Total	$12\,V$		

Next, we simplify the circuit. This means that we calculate the equivalent resistance and redraw the circuit accordingly. We'll first find the equivalent resistance of the parallel part of the circuit:

$$\frac{1}{R_{eq}} = \frac{1}{7\,\Omega} + \frac{1}{10\,\Omega}$$

Use your calculator to get $\dfrac{1}{R_{eq}} = 0.24\,\Omega$.

Taking the reciprocal and rounding to 1 significant figure, we get

$$R_{eq} = 4\ \Omega.$$

So we can redraw our circuit like this:

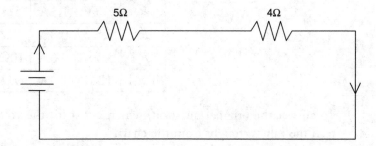

Next, we calculate the equivalent resistance of the entire circuit. Following our rule for resistors in series, we have

$$R_{eq} = 4\ \Omega + 5\ \Omega = 9\ \Omega$$

We can now fill this value into the *V-I-R* chart.

	V	I	R
R_1			5 Ω
R_2			7 Ω
R_3			10 Ω
Total	12 V		9 Ω

Notice that we now have two of the three values in the "Total" row. Using Ohm's law, we can calculate the third. That's the beauty of the *V-I-R* chart: *Ohm's law is valid whenever two of the three entries in a row are known.*

Then we need to put on our thinking caps. We know that all the current that flows through our circuit will also flow through R_1. (You may want to take a look back at the original drawing of our circuit to make sure you understand why this is so.) Therefore, the *I* value in the "R_1" row will be the same as the *I* in the "Total" row. We now have two of the three values in the "R_1" row, so we can solve for the third using Ohm's law.

	V	I	R
R_1	6.5 V	1.3 A	5 Ω
R_2			7 Ω
R_3			10 Ω
Total	12 V	1.3 A	9 Ω

Finally, we know that the voltage across R_2 equals the voltage across R_3, because these resistors are connected in parallel. The total voltage across the circuit is 12 V, and the voltage across R_1 is 6.5 V. So the voltage that occurs between R_1 and the end of the circuit is

$$12\ V - 6.5\ V = 5.5\ V$$

Therefore, the voltage across R_2, which is the same as the voltage across R_3, is 5.5 V. We can fill this value into our table. Finally, we can use Ohm's law to calculate I for both R_2 and R_3. The finished *V-I-R* chart looks like this:

	V	I	R
R_1	6.5 V	1.3 A	5 Ω
R_2	5.5 V	0.8 A	7 Ω
R_3	5.5 V	0.6 A	10 Ω
Total	12 V	1.3 A	9 Ω

To answer the original question, which asked for the voltage across each resistor, we just read the values straight from the chart.

Now, you might be saying to yourself, "This seems like an awful lot of work to solve a relatively simple problem." You're right—it is.

However, there are several advantages to the *V-I-R* chart. The major advantage is that, by using it, you force yourself to approach every circuit problem exactly the same way. So when you're under pressure—as you will be during the AP exam—you'll have a tried-and-true method to turn to.

Also, if there are a whole bunch of resistors, you'll find that the *V-I-R* chart is a great way to organize all your calculations. That way, if you want to check your work, it'll be very easy to do.

Finally, free-response problems that involve circuits generally ask you questions like these.

(a) What is the voltage across each resistor?
(b) What is the current flowing through resistor #4?
(c) What is the power dissipated by resistor #2?

By using the *V-I-R* chart, you do all your calculations once, and then you have all the values you need to solve any question that the AP writers could possibly throw at you.

Tips for Solving Circuit Problems Using the *V-I-R* Chart

- First, enter all the given information into your chart. If resistors haven't already been given names (like " R_1 "), you should name them for easy reference.
- Next simplify the circuit to calculate R_{eq}, if possible.
- Once you have two values in a row, you can calculate the third using Ohm's law. *You CANNOT use Ohm's law unless you have two of the three values in a row.*
- Remember that if two resistors are in series, the current through one of them equals the current through the other. And if two resistors are in parallel, the voltage across one equals the voltage across the other.

Kirchoff's Laws

Kirchoff's laws help you solve complicated circuits. They are especially useful if your circuit contains two batteries.

Kirchoff's laws say:

> 1. At any junction, the current entering equals the current leaving.
> 2. The sum of voltages around a closed loop is 0.

The first law is called the "junction rule," and the second is called the "loop rule." To illustrate the junction rule, we'll revisit the circuit from our first problem. (See Figure 20.6.)

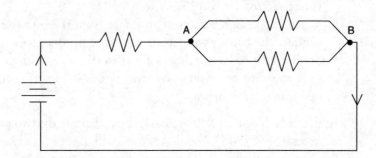

Figure 20.6   Circuit illustrating Kirchoff's junction rule.

According to the junction rule, whatever current enters Junction "A" must also leave Junction "A." So let's say that 1.25 A enters Junction "A," and then that current gets split between the two branches. If we measured the current in the top branch and the current in the bottom branch, we would find that the total current equals 1.25 A. And, in fact, when the two branches came back together at Junction "B," we would find that exactly 1.25 A was flowing out through Junction "B" and through the rest of the circuit.

Kirchoff's junction rule says that charge is conserved: you don't lose any current when the wire bends or branches. This seems remarkably obvious, but it's also remarkably essential to solving circuit problems.

Kirchoff's loop rule is a bit less self-evident, but it's quite useful in sorting out difficult circuits.

As an example, we'll show you how to use Kirchoff's loop rule to find the current through all the resistors in the circuit below.

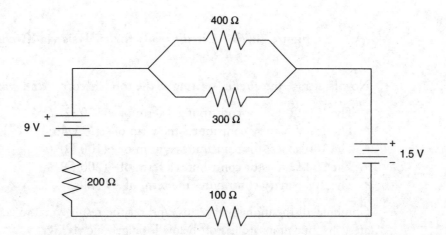

Figure 20.7a   Example circuit for using Kirchoff's loop rule.

We will follow the steps for using Kirchoff's loop rule (Figure 20.7a):

• Arbitrarily choose a direction of current. Draw arrows on your circuit to indicate this direction.

- Follow the loop in the direction you chose. When you cross a resistor, the voltage is $-IR$, where R is the resistance, and I is the current flowing through the resistor. This is just an application of Ohm's law. (If you have to follow a loop *against* the current, though, the voltage across a resistor is written $+IR$.)
- When you cross a battery, if you trace from the − to the + add the voltage of the battery, subtract the battery's voltage if you trace from + to −.
- Set the sum of your voltages equal to 0. Solve. If the current you calculate is negative, then the direction you chose was wrong—the current actually flows in the direction opposite to your arrows.

In the case of Figure 20.7b, we'll start by collapsing the two parallel resistors into a single equivalent resistor of 170 Ω. You don't *have* to do this, but it makes the mathematics much simpler.

Next, we'll choose a direction of current flow. But which way? In this particular case, you can probably guess that the 9 V battery will dominate the 1.5 V battery, and thus the current will be clockwise. But even if you aren't sure, just choose a direction and stick with it—if you get a negative current, you chose the wrong direction.

Here is the circuit redrawn with the parallel resistors collapsed and the assumed direction of current shown. Because there's now only one path for current to flow through, we have labeled that current I.

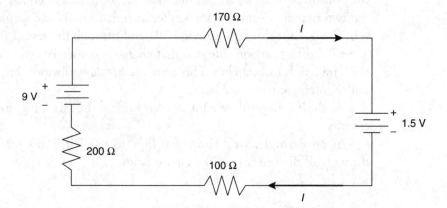

Figure 20.7b Circuit ready for analysis via Kirchoff's loop rule.

Now let's trace the circuit, starting at the top left corner and working clockwise:

- The 170 Ω resistor contributes a term of $-(170\ \Omega)\ I$.
- The 1.5 V battery contributes the term of −1.5 volts.
- The 100 Ω resistor contributes a term of $-(100\ \Omega)\ I$.
- The 200 Ω resistor contributes a term of $-(200\ \Omega)\ I$.
- The 9 V battery contributes the term of +9 volts.

Combine all the individual terms, and set the result equal to zero. The units of each term are volts, but units are left off below for algebraic clarity:

$$0 = (-170)I + (-1.5) + (-100)I + (-200)I + (+9)$$

By solving for I, the current in the circuit is found to be 0.016 A; that is, 16 milliamps, a typical laboratory current.

The problem is not yet completely solved, though—16 milliamps go through the 100 Ω and 200 Ω resistors, but what about the 300 Ω and 400 Ω resistors? We can find that the voltage across the 170 Ω equivalent resistance is (0.016 A)(170 Ω) = 2.7 V. Because the voltage across parallel resistors is the same for each, the current through each is just 2.7 V divided by the resistance of the actual resistor: 2.7 V/300 Ω = 9 mA, and 2.7 V/400 Ω = 7 mA. Problem solved!

Oh, and you might notice that the 9 mA and 7 mA through each of the parallel branches adds to the total of 16 mA—as required by Kirchoff's junction rule.

Circuits from an Experimental Point of View

When a real circuit is set up in the laboratory, it usually consists of more than just resistors—light bulbs and motors are common devices to hook to a battery, for example. For the purposes of computation, though, we can consider pretty much any electronic device to act like a resistor.

But what if your purpose is *not* computation? Often on the AP exam, as in the laboratory, you are asked about observational and measurable effects. The most common questions involve the brightness of light bulbs and the measurement (not just computation) of current and voltage.

Brightness of a Bulb

The brightness of a bulb depends solely on the power dissipated by the bulb. (Remember, power is given by any of the equations I^2R, IV, or V^2/R). You can remember that from your own experience—when you go to the store to buy a light bulb, you don't ask for a "400-ohm" bulb, but for a "100-watt" bulb. And a 100-watt bulb is brighter than a 25-watt bulb. But be careful—a bulb's power can change depending on the current and voltage it's hooked up to. Consider this problem.

A light bulb is rated at 100 W in the United States, where the standard wall outlet voltage is 120 V. If this bulb were plugged in in Europe, where the standard wall outlet voltage is 240 V, which of the following would be true?

(A) The bulb would be one-quarter as bright.
(B) The bulb would be one-half as bright.
(C) The bulb's brightness would be the same.
(D) The bulb would be twice as bright.
(E) The bulb would be four times as bright.

Your first instinct might be to say that because brightness depends on power, the bulb is exactly as bright. But that's not right! The power of a bulb can change.

The resistance of a light bulb is a property of the bulb itself, and so will not change no matter what the bulb is hooked to.

Since the resistance of the bulb stays the same while the voltage changes, by V^2/R, the power goes up, and the bulb will be brighter. How much brighter? Since the voltage in Europe is doubled, and because voltage is squared in the equation, the power is multiplied by 4—choice E.

Ammeters and Voltmeters

Ammeters measure current, and voltmeters measure voltage. This is pretty obvious, because current is measured in amps, voltage in volts. It is *not* necessarily obvious, though, how to connect these meters into a circuit.

Remind yourself of the properties of series and parallel resistors—voltage is the same for any resistors in parallel with each other. So if you're going to measure the voltage across a resistor, you must put the voltmeter in *parallel* with the resistor. In Figure 20.8, the meter labeled V_2 measures the voltage across the 100 Ω resistor, while the meter labeled V_1 measures the potential difference between points A and B (which is also the voltage across R_1).

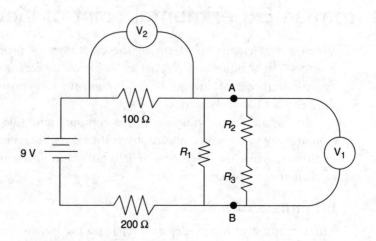

Figure 20.8   Measuring voltage with a voltmeter.

Current is the same for any resistors in *series* with one another. So, if you're going to measure the current through a resistor, the ammeter must be in series with that resistor. In Figure 20.9, ammeter A_1 measures the current through resistor R_1, while ammeter A_2 measures the current through resistor R_2.

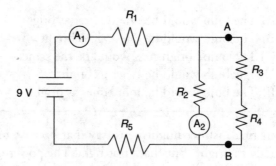

Figure 20.9   Measuring current with an ammeter.

As an exercise, ask yourself, is there a way to figure out the current in the other three resistors based only on the readings in these two ammeters? Answer is in the footnote.[2]

---

[2] The current through R_5 must be the same as through R_1, because both resistors carry whatever current came directly from the battery. The current through R_3 and R_4 can be determined from Kirchoff's junction rule: subtract the current in R_2 from the current in R_1 and that's what's left over for the right-hand branch of the circuit.

RC Circuits: Steady-State Behavior

When you have both resistors and capacitors in a circuit, the circuit is called an "RC circuit." If you remember, we introduced capacitors in Chapter 20, when we talked about charged, parallel plates.

The only situation in which capacitors will show up in a circuit is when the circuit exhibits "steady-state behavior": this just means that the circuit has been connected for a while. In these cases, the only thing you'll generally need to worry about is how to deal with capacitors in series and in parallel.

When capacitors occur in series, you add them inversely. The charge stored on each capacitor in series must be the same.

$$\frac{1}{C_{eq}} = \frac{1}{C_1} + \frac{1}{C_2} + \frac{1}{C_3} + \ldots$$

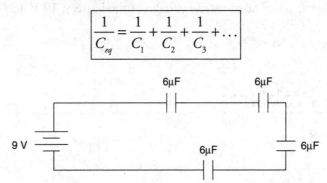

Figure 20.10   Example of capacitors in series.

For the circuit in Figure 20.10, the equivalent capacitance is $C_{eq} = 1.5 \ \mu F$.

When capacitors occur in parallel, you add them algebraically. The voltage across each capacitor in parallel must be the same.

$$C_{eq} = C_1 + C_2 + C_3 + \ldots$$

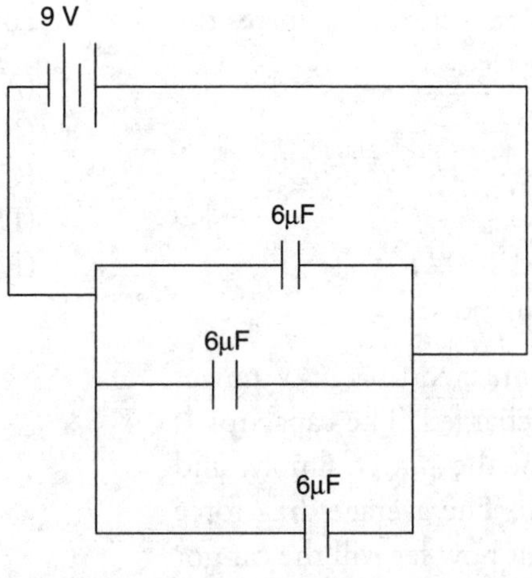

9 V

6μF

6μF

6μF

Figure 20.11   Example of capacitors in parallel.

The equivalent capacitance for the circuit in Figure 20.11 is 18 μF.

You should also know that the energy stored by a capacitor is

$$E = \tfrac{1}{2} \, CV^2$$

Once the circuit has been connected for a long time, capacitors stop current from flowing. To find the charge stored on or the voltage across a capacitor, just use the equation for capacitors, $Q = CV$.

For example, imagine that you hook up a 10-V battery to a 5 Ω resistor in series with an uncharged 1 F capacitor. (1 F capacitors are rarely used in actual electronics application—most capacitances are micro- or nanofarads—but they are commonly used for physics class demonstrations!) When the circuit is first hooked up, the capacitor is empty—it is ready and waiting for as much charge as can flow to it. Thus, initially, the circuit behaves as if the capacitor weren't there. In this case, then, the current through the resistor starts out at 10 V/5 Ω = 2 A.

But, after a long time, the capacitor blocks current. The resistor might as well not be there; we might as well just have a capacitor right across the battery. After a long time, the capacitor takes on the voltage of the battery, 10 V. (So the charge stored on the capacitor is $Q = CV = 10$ C.)

› Practice Problems

Multiple Choice:

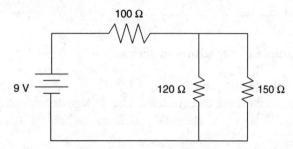

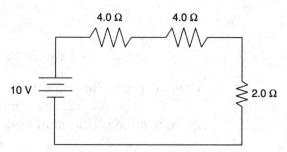

1. A 100 Ω, 120 Ω, and 150 Ω resistor are connected to a 9-V battery in the circuit shown above. Which of the three resistors dissipates the most power?

   (A) the 100 Ω resistor
   (B) the 120 Ω resistor
   (C) the 150 Ω resistor
   (D) both the 120 Ω and 150 Ω
   (E) all dissipate the same power

2. A 1.0-F capacitor is connected to a 12-V power supply until it is fully charged. The capacitor is then disconnected from the power supply, and used to power a toy car. The average drag force on this car is 2 N. About how far will the car go?

   (A) 36 m
   (B) 72 m
   (C) 144 m
   (D) 24 m
   (E) 12 m

3. Three resistors are connected to a 10-V battery as shown in the diagram above. What is the current through the 2.0 Ω resistor?

   (A) 0.25 A
   (B) 0.50 A
   (C) 1.0 A
   (D) 2.0 A
   (E) 4.0 A

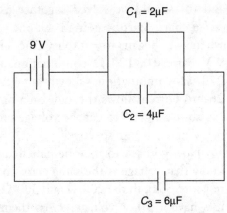

$C_1 = 2\mu F$

9 V

$C_2 = 4\mu F$

$C_3 = 6\mu F$

4. Three capacitors are connected as shown in the diagram above. $C_1 = 2\mu F$; $C_2 = 4\mu F$; $C_3 = 6\mu F$. If the battery provides a potential of 9 V, how much charge is stored by this system of capacitors?

(A) 3.0 μC
(B) 30 μC
(C) 2.7 μC
(D) 27 μC
(E) 10 μC

5. What is the resistance of an ideal ammeter and an ideal voltmeter?

Ideal Ammeter	Ideal Voltmeter
(A) zero	infinite
(B) infinite	zero
(C) zero	zero
(D) infinite	infinite
(E) 1 Ω	1 Ω

Free Response:

6.

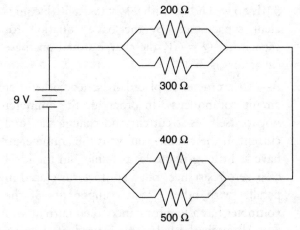

200 Ω

300 Ω

9 V

400 Ω

500 Ω

(a) Simplify the above circuit so that it consists of one equivalent resistor and the battery.

(b) What is the total current through this circuit?

(c) Find the voltage across each resistor. Record your answers in the spaces below.

Voltage across 200 Ω resistor: _____
Voltage across 300 Ω resistor: _____
Voltage across 400 Ω resistor: _____
Voltage across 500 Ω resistor: _____

(d) Find the current through each resistor. Record your answers in the spaces below.
Current through 200 Ω resistor: _____
Current through 300 Ω resistor: _____
Current through 400 Ω resistor: _____
Current through 500 Ω resistor: _____

(e) The 500 Ω resistor is now removed from the circuit. State whether the current through the 200 Ω resistor would increase, decrease, or remain the same. Justify your answer.

› Solutions to Practice Problems

1. **A**—On one hand, you could use a *V-I-R* chart to calculate the voltage or current for each resistor, then use $P = IV$, I^2R, or V^2/R to find power. On the other hand, there's a quick way to reason through this one. Voltage changes across the 100 Ω resistor, then again across the parallel combination. Because the 100 Ω resistor has a bigger resistance than the parallel combination, the voltage across it is larger as well. Now consider each resistor individually. By power $= V^2/R$, the 100 Ω resistor has both the biggest voltage and the smallest resistance, giving it the most power.

2. **A**—The energy stored by a capacitor is $\frac{1}{2}CV^2$. By powering a car, this electrical energy is converted into mechanical work, equal to force times parallel displacement. Solve for displacement, you get 36 m.

3. **C**—To use Ohm's law here, simplify the circuit to a 10 V battery with the 10 Ω equivalent resistance. We can use Ohm's law for the entire circuit to find that 1.0 A is the total current. Because all the resistors are in series, this 1.0 A flows through each resistor, including the 2 Ω resistor.

4. D—First, simplify the circuit to find the equivalent capacitance. The parallel capacitors add to 6 μF. Then the two series capacitors combine to 3 μF. So we end up with 9 V across a 3-μF equivalent capacitance. By the basic equation for capacitors, $Q = CV$, the charge stored on these capacitors is 27 μC.

5. A—An ammeter is placed in series with other circuit components. In order for the ammeter not to itself resist current and change the total current in the circuit, you want the ammeter to have as little resistance as possible—in the ideal case, zero resistance. But a voltmeter is placed in parallel with other circuit components. If the voltmeter has a low resistance, then current will flow through the voltmeter instead of through the rest of the circuit. Therefore, you want it to have as high a resistance as possible, so the voltmeter won't affect the circuit being measured.

6. (a) Combine each of the sets of parallel resistors first. You get 120 Ω for the first set, 222 Ω for the second set, as shown in the diagram below. These two equivalent resistances add as series resistors to get a total resistance of 342 Ω.

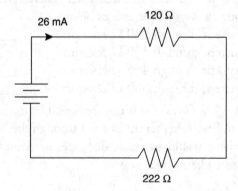

26 mA   120 Ω

222 Ω

(b) Now that we've found the total resistance and we were given the total voltage, just use Ohm's law to find the total current to be 0.026 A (also known as 26 mA).

(c) and (d) should be solved together using a *V-I-R* chart. Start by going back one step to when we began to simplify the circuit: 9-V battery, a 120 Ω combination, and a 222 Ω combination, shown above. The 26-mA current flows through each of these . . . so use $V = IR$ to get the voltage of each: 3.1 V and 5.8 V, respectively.

Now go back to the original circuit. We know that voltage is the same across parallel resistors. So both the 200-Ω and 300-Ω resistors have a 3.1-V voltage across them. Use Ohm's law to find that 16 mA goes through the 200-Ω resistor, and 10 mA through the 300 Ω. Similarly, both the 400-Ω and 500-Ω resistors must have 5.8 V across them. We get 15 mA and 12 mA, respectively.

Checking these answers for reasonability: the total voltage adds to 8.9 V, or close enough to 9.0 V with rounding. The current through each set of parallel resistors adds to just about 26 mA, as we expect.

(e) Start by looking at the circuit as a whole. When we remove the 500 Ω resistor, we actually *increase* the overall resistance of the circuit because we have made it more difficult for current to flow by removing a parallel path. The total voltage of the circuit is provided by the battery, which provides 9.0 V no matter what it's hooked up to. So by Ohm's law, if total voltage stays the same while total resistance increases, total current must *decrease* from 26 mA.

Okay, now look at the first set of parallel resistors. Their equivalent resistance doesn't change, yet the total current running through them decreases, as discussed above. Therefore, the voltage across each resistor decreases, and the current through each decreases as well.

❯ Rapid Review

- Current is the flow of positive charge. It is measured in amperes.

- Resistance is a property that impedes the flow of charge. Resistance in a circuit comes from the internal resistance of the wires and from special elements inserted into circuits known as "resistors."

- Resistance is related to current and voltage by Ohm's law: $V = IR$.

- When resistors are connected in series, the total resistance equals the sum of the individual resistances. And the current through one resistor equals the current through any other resistor in series with it.

- When resistors are connected in parallel, the inverse of the total resistance equals the sum of the inverses of the individual resistances. The voltage across one resistor equals the voltage across any other resistor connected parallel to it.

Exam tip from an AP Physics veteran:

Many AP problems test your ability to use Ohm's law correctly. Ohm's law cannot be used unless the voltage, current, and resistance all refer to the same circuit element; on a *V-I-R* chart, this means that Ohm's law can only be used across a single row of the chart.

—*Chat, college junior and physics major*

- The *V-I-R* chart is a convenient way to organize any circuit problem.

- Kirchoff's junction rule says that any current coming into a junction will leave the junction. This is a statement of conservation of charge. Kirchoff's loop rule says that the sum of the voltages across a closed loop equals zero. This rule is helpful especially when solving problems with circuits that contain more than one battery.

- Ammeters measure current, and are connected in series; voltmeters measure voltage, and are connected in parallel.

- When capacitors are connected in series, the inverse of the total capacitance equals the sum of the inverses of the individual capacitances. When they are connected in parallel, the total capacitance just equals the sum of the individual capacitances.

- A capacitor's purpose in a circuit is to store charge. After it has been connected to a circuit for a long time, the capacitor becomes fully charged and prevents the flow of current.

CHAPTER 21

Magnetism

IN THIS CHAPTER

Summary: Magnetic fields produce forces on moving charges; moving charges, such as current-carrying wires, can create magnetic fields. This chapter discusses the production and the effects of magnetic fields.

Key Ideas

- ✪ The force on a moving charge due to a magnetic field is qvB.
- ✪ The direction of the magnetic force on a moving charge is given by a right hand rule, and is NOT in the direction of the magnetic field.
- ✪ Current-carrying wires produce magnetic fields.
- ✪ When the magnetic flux through a wire changes, a voltage is induced.

Relevant Equations

Force on a charged particle in a magnetic field:

$$F = qvB$$

Force on a current-carrying wire:

$$F = ILB$$

Magnetic field due to a long, straight, current-carrying wire:

$$B = \frac{\mu_0 I}{2\pi r}$$

Magnetic flux:

$$\Phi_B = BA$$

Induced EMF:

$$\varepsilon = N\frac{\Delta\Phi}{\Delta t}$$

Induced EMF for a rectangular wire moving into or out of a magnetic field:

$$\varepsilon = BLv$$

When most people think of magnets, they imagine horseshoe-shaped objects that can pick up bits of metal. Or maybe they visualize a refrigerator door. But not a physics ace like you! You know that magnetism is a wildly diverse topic, involving everything from bar magnets to metal coils to mass spectrometers. Perhaps you also know that magnetism is a subject filled with countless "right-hand rules," many of which can seem difficult to use or just downright confusing. So our goal in this chapter—besides reviewing all of the essential concepts and formulas that pertain to magnetism—is to give you a set of easy to understand, easy to use right-hand rules that are guaranteed to earn you points on the AP exam.

Magnetic Fields

All magnets are dipoles, which means that they have two "poles," or ends. One is called the north pole, and the other is the south pole. Opposite poles attract, and like poles repel.

You can never create a magnet with just a north pole or just a south pole. If you took the magnet in Figure 21.1

Figure 21.1   Bar magnet.

and cut in down the middle, you would not separate the poles. Instead, you would create two magnets like those shown in Figure 21.2.

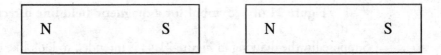

Figure 21.2   Cutting the bar magnet in Figure 21.1 in half just gives you two smaller bar magnets. You can never get an isolated north or south pole.

A magnet creates a magnetic field. (See Figure 21.3.) Unlike electric field lines, which either go from a positive charge to a negative charge or extend infinitely into space, magnetic field lines form loops. These loops point away from the north end of a magnet, and toward the south end. Near the magnet the lines point nearly straight into or out of the pole.

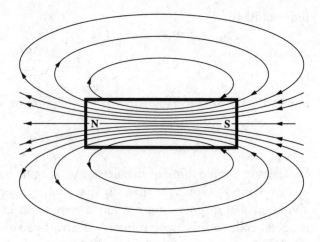

Figure 21.3   Magnetic field lines created by a bar magnet.

Just as we talk about the value of an electric field at a certain point, we can also talk about the value of a magnetic field at a certain point. The value of a magnetic field is a vector quantity, and it is abbreviated with the letter B. The value of a magnetic field is measured in teslas.

Often, the writers of the AP exam like to get funky about how they draw magnetic field lines. Rather than putting a magnetic field in the plane of the page, so that the field would point up or down or left or right, the AP writers will put magnetic fields perpendicular to the page. This means that the magnetic field either shoots out toward you or shoots down into the page.

When a magnetic field line is directed out of the page, it is drawn as shown in Figure 21.4a,

Figure 21.4a   Symbol for a magnetic field line directed out of the page.

and when a magnetic field line is directed into the page, it is drawn as shown in Figure 21.4b.

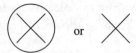

Figure 21.4b   Symbol for a magnetic field line directed into the page.

Supposedly, the drawing in Figure 21.4a is intended to look like the tip of an arrow coming out of a page, and the drawing in Figure 21.4b is intended to look like the tail of an arrow going into a page.[1] These symbols can be used to describe other ideas, such as electric fields going into or out of the page, or currents flowing into or out of the page, but they are most often used to describe magnetic fields.

---

[1]If you're not too impressed by these representations, just remember how physicists like to draw refrigerators. There's a reason why these science folks weren't accepted into art school.

Long, Straight, Current-Carrying Wires

Bar magnets aren't the only things that create magnetic fields—current-carrying wires do also. Of course, you can also create a magnetic field using a short, curvy, current-carrying wire, but the equations that describe that situation are a little more complicated, so we'll focus on long, straight, current-carrying wires.

The magnetic field created by a long, straight, current-carrying wire loops around the wire in concentric circles. The direction in which the magnetic field lines loop is determined by a right-hand rule.

(Incidentally, our versions of the right-hand rules may not be the same as what you've learned in physics class. If you're happy with the ones you already know, you should ignore our advice and just stick with what works best for you.)

Right-hand rule: To find the direction of the B field produced by long, straight, current-carrying wires.

Pretend you are holding the wire with your right hand. Point your thumb in the direction of the current. Your fingers wrap around your thumb the same way that the magnetic field wraps around the wire.

Here's an example. A wire is directed perpendicular to the plane of this page (that is, it's coming out straight toward you). The current in this wire is flowing out of the page. What does the magnetic field look like?

To solve this, we first pretend that we are grabbing the wire. If it helps, take your pencil and place it on this page, with the eraser touching the page and the point of the pencil coming out toward you. This pencil is like the wire. Now grab the pencil with your right hand. The current is coming out of the page, so make sure that you have grabbed the pencil in such a way that your thumb is pointing away from the page. If it looks like you're giving someone a "thumbs-up sign," then you're doing this correctly. Finally, look at how your fingers are wrapped around the pencil. From a birds-eye view, it should look like your fingers are wrapping counterclockwise. So this tells us the answer to the problem, as shown in Figure 21.5.

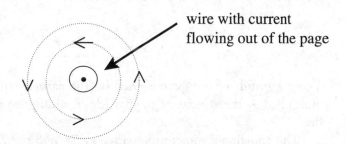

Figure 21.5   Magnetic field (dotted lines) generated by a long, straight, current-carrying wire oriented perpendicular to the plane of the page.

Here's another example. What does the magnetic field look like around a wire in the plane of the page with current directed upward?

We won't walk you through this one; just use the right-hand rule, and you'll be fine. The answer is shown in Figure 21.6.

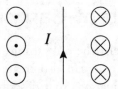

**Figure 21.6   Magnetic field around a wire in the
plane of the page with current directed upward.**

The formula that describes the magnitude of the magnetic field created by a long, straight, current-carrying wire is the following:

$$B = \frac{\mu_0 I}{2\pi r}$$

In this formula, B is the magnitude of the magnetic field, μ_0 is a constant called the "permeability of free space" ($\mu_0 = 4\pi \times 10^{-7}$ T·m/A), I is the current flowing in the wire, and r is the distance from the wire.

Moving Charged Particles

The whole point of defining a magnetic field is to determine the forces produced on an object by the field. You are familiar with the forces produced by bar magnets—like poles repel, opposite poles attract. We don't have any formulas for the amount of force produced in this case, but that's okay, because this kind of force is irrelevant to the AP exam.

Instead, we must focus on the forces produced by magnetic fields on charged particles, including both isolated charges and current-carrying wires. (After all, current is just the movement of positive charges.)

A magnetic field exerts a force on a charged particle if that particle is moving perpendicular to the magnetic field. A magnetic field does not exert a force on a stationary charged particle, nor on a particle that is moving parallel to the magnetic field.

$$F = qvB$$

The magnitude of the force exerted on the particle equals the charge on the particle, q, multiplied by the velocity of the particle, v, multiplied by the magnitude of the magnetic field.

This equation is sometimes written as $F = qvB(\sin \theta)$. The θ refers to the angle formed between the velocity vector of your particle and the direction of the magnetic field. So, if a particle moves in the same direction as the magnetic field lines, $\theta = 0°$, $\sin 0° = 0$, and *that particle experiences no magnetic force!*

Nine times out of ten, you will not need to worry about this "sin θ" term, because the angle will either be zero or 90°. However, if a problem explicitly tells you that your particle is *not* traveling perpendicular to the magnetic field, then you will need to throw in this extra "sin θ" term.

Right-hand rule: To find the force on a charged particle.

   Point your right hand, with fingers extended, in the direction that the charged particle is traveling. Then, bend your fingers so that they point in the direction of the magnetic field.

- If the particle has a POSITIVE charge, your thumb points in the direction of the force exerted on it.
- If the particle has a NEGATIVE charge, your thumb points opposite the direction of the force exerted on it.

The key to this right-hand rule is to remember the sign of your particle. This next problem illustrates how important sign can be.

An electron travels through a magnetic field, as shown below. The particle's initial velocity is 5×10^6 m/s, and the magnitude of the magnetic field is 0.4 T. What are the magnitude and direction of the particle's acceleration?

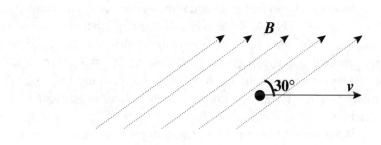

This is one of those problems where you're told that the particle is *not* moving perpendicular to the magnetic field. So the formula we use to find the magnitude of the force acting on the particle is

$$F = qvB(\sin \theta)$$
$$F = (1.6 \times 10^{-19} \text{ C})(5 \times 10^6 \text{ m/s})(0.4 \text{ T})(\sin 30°)$$
$$F = 1.6 \times 10^{-13} \text{ N}$$

Note that we never plug in the negative signs when calculating force. The negative charge on an electron will influence the direction of the force, which we will determine in a moment. Now we solve for acceleration:

$$F_{net} = ma$$
$$a = 1.8 \times 10^{17} \text{ m/s}^2$$

Wow, you say . . . a bigger acceleration than anything we've ever dealt with. Is this unreasonable? After all, in less than a second the particle would be moving faster than the speed of light, right? The answer is still reasonable. In this case, the acceleration is perpendicular to the velocity. This means the acceleration is *centripetal*, and the particle must move in a circle at constant speed. But even if the particle were speeding up at this rate, either the acceleration wouldn't act for very long, or relativistic effects would prevent the particle from traveling faster than light.

Finally, we solve for direction using the right-hand rule. We point our hand in the direction that the particle is traveling—to the right. Next, we curl our fingers upward, so that they point in the same direction as the magnetic field. Our thumb points out of the page. BUT WAIT!!! We're dealing with an electron, which has a negative charge. So the force acting on our particle, and therefore the particle's acceleration, points in the opposite direction. The particle is accelerating into the page.

Magnetic Force on a Wire

A current is simply the flow of positive charges. So, if we put a current-carrying wire perpendicular to a magnetic field, we have placed moving charges perpendicular to the field, and these charges experience a force. The wire can be pulled by the magnetic field!

The formula for the force on a long, straight, current-carrying wire in the presence of a magnetic field is

$$F = ILB$$

This equation says that the force on a wire equals the current in the wire, I, multiplied by the length of the wire, L, multiplied by the magnitude of the magnetic field, B, in which the wire is located.

Sometimes you'll see this equation written as $F = ILB(\sin \theta)$. Just like the equation for the force on a charge, the θ refers to the angle between the wire and the magnetic field. You normally don't have to worry about this θ because, in most problems, the wire is perpendicular to the magnetic field, and $\sin 90° = 1$, so the term cancels out.

The direction of the force on a current-carrying wire is given by the same right-hand rule as for the force on a charged particle because current is simply the flow of positive charge.

What would happen if you had two long, straight, current-carrying wires side by side? This is a question that the writers of the AP exam love to ask, so it is a great idea to learn how to answer it.

The trick that makes answering this question very easy is that you have to draw the direction of the magnetic field that one of the wires creates; then consider the force on the other wire. So, for example . . .

Two wires are placed parallel to each other. The direction of current in each wire is indicated above. How will these wires interact?

(A) They will attract each other.
(B) They will repel each other.
(C) They will not affect each other.
(D) This question cannot be answered without knowing the length of each wire.
(E) This question cannot be answered without knowing the current in each wire.

Let's follow our advice and draw the magnetic field created by the left-hand wire.

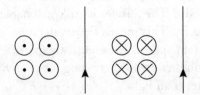

Now, a wire's field cannot produce a force on itself. The field that we drew is *caused by* the left wire, but produces a force on the right-hand wire. Which direction is that force? Use the right-hand rule for the force on a charged particle. The charges are moving up, in the direction of the current. So point up the page, and curl your fingers toward the magnetic field, into the page. The right wire is forced to the LEFT. Newton's third law says that the force on the left wire by the right wire will be equal and opposite.[2] So, the wires attract, answer A.

Often, textbooks give you advice such as, "Whenever the current in two parallel wires is traveling in the same direction, the wires will attract each other, and vice versa." Use it if you like, but this advice can easily be confused.

Mass Spectrometry: More Charges Moving Through Magnetic Fields

A magnetic field can make a charged particle travel in a circle. Here's how it performs this trick.

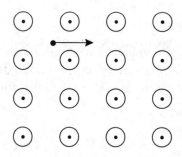

Figure 21.7a Positively charged particle moving in a magnetic field directed out of the page.

Let's say you have a proton traveling through a uniform magnetic field coming out of the page, and the proton is moving to the right, like the one we drew in Figure 21.7a. The magnetic field exerts a downward force on the particle (use the right-hand rule). So the path of the particle begins to bend downward, as shown in Figure 21.7b.

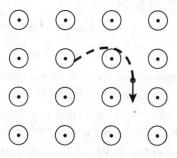

Figure 21.7b Curving path of a positively charged particle moving in a magnetic field directed out of the page.

---

[2] You could also figure out the force on the left wire by using the same method we just used for the force on the right wire: draw the magnetic field produced by the right wire, and use the right-hand rule to find the direction of the magnetic force acting on the left wire.

Now our proton is moving straight down. The force exerted on it by the magnetic field, using the right-hand rule, is now directed to the left. So the proton will begin to bend leftward. You probably see where this is going—a charged particle, traveling perpendicular to a uniform magnetic field, will follow a circular path.

We can figure out the radius of this path with some basic math. The force of the magnetic field is causing the particle to go in a circle, so this force must cause centripetal acceleration. That is,

$$qvB = \frac{mv^2}{r}$$

We didn't include the "sin θ" term because the particle is always traveling perpendicular to the magnetic field. We can now solve for the radius of the particle's path:

$$r = \frac{mv}{qB}$$

The real-world application of this particle-in-a-circle trick is called a mass spectrometer. A mass spectrometer is a device used to determine the mass of a particle.

A mass spectrometer, in simplified form, is drawn in Figure 21.8.

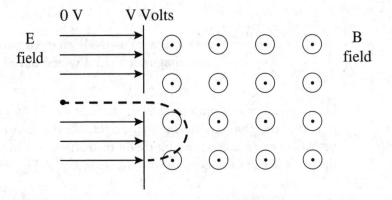

Figure 21.8   Basic mass spectrometer.

A charged particle enters a uniform electric field (shown at the left in Figure 21.8). It is accelerated by the electric field. By the time it gets to the end of the electric field, it has acquired a high velocity, which can be calculated using conservation of energy. Then the particle travels through a tiny opening and enters a uniform magnetic field. This magnetic field exerts a force on the particle, and the particle begins to travel in a circle. It eventually hits the wall that divides the electric-field region from the magnetic-field region. By measuring where on the wall it hits, you can determine the radius of the particle's path. Plugging this value into the equation above that we derived for the radius of the path, you can calculate the particle's mass.

You may see a problem on the free-response section that involves a mass spectrometer. These problems may seem intimidating, but, when you take them one step at a time, they're not very difficult.

Induced EMF

A changing magnetic field produces a current. We call this occurrence **electromagnetic induction**.

Let's say that you have a loop of wire in a magnetic field. Under normal conditions, no current flows in your wire loop. However, if you change the magnitude of the magnetic field, a current will begin to flow.

We've said in the past that current flows in a circuit (and a wire loop qualifies as a circuit, albeit a simple one) when there is a potential difference between the two ends of the circuit. Usually, we need a battery to create this potential difference. But we don't have a battery hooked up to our loop of wire. Instead, the changing magnetic field is doing the same thing as a battery would. So rather than talking about the voltage of the battery in this circuit, we talk about the "voltage" created by the changing magnetic field. The technical term for this "voltage" is **induced EMF**.

> **Induced EMF:** The potential difference created by a changing magnetic field that causes a current to flow in a wire. EMF stands for Electro-Motive Force, but is *NOT* a force.

For a loop of wire to "feel" the changing magnetic field, some of the field lines need to pass through it. The amount of magnetic field that passes through the loop is called the **magnetic flux**. This concept is pretty similar to electric flux.

> **Magnetic Flux:** The number of magnetic field lines that pass through an area

The units of flux are called webers; 1 weber = 1 T·m². The equation for magnetic flux is

$$\Phi_B = BA$$

In this equation, Φ_B is the magnetic flux, B is the magnitude of the magnetic field, and A is the area of the region that is penetrated by the magnetic field.

Let's take a circular loop of wire, lay it down on the page, and create a magnetic field that points to the right, as shown in Figure 21.9.

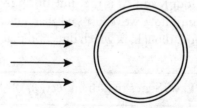

Figure 21.9   Loop of wire in the plane of a magnetic field.

No field lines go through the loop. Rather, they all hit the edge of the loop, but none of them actually passes through the center of the loop. So we know that our flux should equal zero.

Okay, this time we will orient the field lines so that they pass through the middle of the loop. We'll also specify the loop's radius = 0.2 m, and that the magnetic field is that of the Earth, $B = 5 \times 10^{-5}$ T. This situation is shown in Figure 21.10.

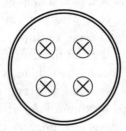

Figure 21.10   Loop of wire with magnetic field lines going through it.

Now all of the area of the loop is penetrated by the magnetic field, so A in the flux formula is just the area of the circle, πr^2.

The flux here is

$$\Phi_B = (5 \times 10^{-5})(\pi)(0.2^2) = 6.2 \times 10^{-6} \text{ T·m}^2$$

Sometimes you'll see the flux equation written as $BA\cos\theta$. The additional cosine term is only relevant when a magnetic field penetrates a wire loop at some angle that's not 90°. The angle θ is measured between the magnetic field and the "normal" to the loop of wire . . . if you didn't get that last statement, don't worry about it. Rather, know that the cosine term goes to 1 when the magnetic field penetrates directly into the loop, and the cosine term goes to zero when the magnetic field can't penetrate the loop at all.

Because a loop will only "feel" a changing magnetic field if some of the field lines pass through the loop, we can more accurately say the following: *A changing magnetic **flux** creates an induced EMF.*

Faraday's law tells us exactly how much EMF is induced by a changing magnetic flux.

$$\varepsilon = \frac{N \cdot \Delta \Phi_B}{\Delta t} \Big|^3$$

ε is the induced EMF, N is the number of loops you have (in all of our examples, we've only had one loop), and Δt is the time during which your magnetic flux, Φ_B, is changing.

Up until now, we've just said that a changing magnetic flux creates a current. We haven't yet told you, though, in which direction that current flows. To do this, we'll turn to **Lenz's Law**.

> **Lenz's Law:** States that the direction of the induced current opposes the increase in flux

When a current flows through a loop, that current creates a magnetic field. So what Lenz said is that the current that is induced will flow in such a way that the magnetic field it creates points opposite to the direction in which the already existing magnetic flux is changing.

---

[3]The equation sheet puts a − sign in front of the expression for induced EMF. This is irrelevant for Physics B students.

Sound confusing?[4] It'll help if we draw some good illustrations. So here is Lenz's Law in pictures.

We'll start with a loop of wire that is next to a region containing a magnetic field (Figure 21.11a). Initially, the magnetic flux through the loop is zero.

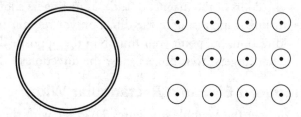

Figure 21.11a   Loop of wire next to a region containing a magnetic field pointing out of the page.

Now, we will move the wire into the magnetic field. When we move the loop toward the right, the magnetic flux will increase as more and more field lines begin to pass through the loop. The magnetic flux is increasing out of the page—at first, there was no flux out of the page, but now there is some flux out of the page. Lenz's Law says that the induced current will create a magnetic field that opposes this increase in flux. So the induced current will create a magnetic field into the page. By the right-hand rule, the current will flow clockwise. This situation is shown in Figure 21.11b.

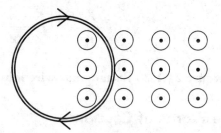

Figure 22.11b   Current induced in loop of wire as it moves into a magnetic field directed out of the page.

After a while, the loop will be entirely in the region containing the magnetic field. Once it enters this region, there will no longer be a changing flux, because no matter where it is within the region, the same number of field lines will always be passing through the loop. Without a changing flux, there will be no induced EMF, so the current will stop. This is shown in Figure 21.11c.

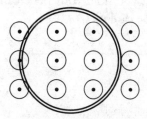

Figure 21.11c   Loop of wire with no current flowing, because it is not experiencing a changing magnetic flux.

---

[4]"Yes !"

To solve a problem that involves Lenz's Law, use this method:

- Point your right thumb in the initial direction of the magnetic field.
- Ask yourself, "Is the flux increasing or decreasing?"
- If the flux is decreasing, then just curl your fingers (with your thumb still pointed in the direction of the magnetic field). Your fingers show the direction of the induced current.
- If flux is increasing in the direction you're pointing, then flux is decreasing in the other direction. So, point your thumb in the opposite direction of the magnetic field, and curl your fingers. Your fingers show the direction of the induced current.

Induced EMF in a Rectangular Wire

Consider the example in Figures 21.11a–c with the circular wire being pulled through the uniform magnetic field. It can be shown that if instead we pull a *rectangular* wire into or out of a uniform field B at constant speed v, then the induced EMF in the wire is found by

$$\varepsilon = BLv$$

Here, L represents the length of the side of the rectangle that is NOT entering or exiting the field, as shown below in Figure 21.12.

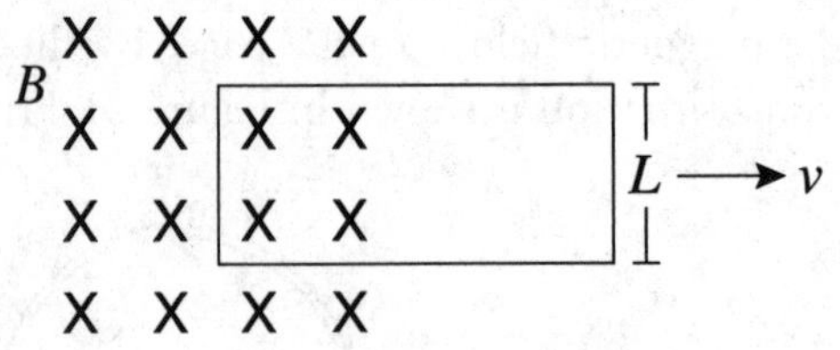

Figure 21.12   Rectangular wire moving through a uniform magnetic field.

Some Words of Caution

We say this from personal experience. First, when using a right-hand rule, use big, easy-to-see gestures. A right-hand rule is like a form of advertisement: it is a way that your hand tells your brain what the answer to a problem is. You want that advertisement to be like a billboard—big, legible, and impossible to misread. Tiny gestures will only lead to mistakes. Second, when using a right-hand rule, *always* use your right hand. *Never use your left hand!* This will cost you points!

Exam tip from an AP Physics veteran:
Especially if you hold your pencil in your right hand, it's easy accidentally to use your left hand. Be careful!

—*Jessica, college sophomore*

〉 Practice Problems

Multiple Choice:

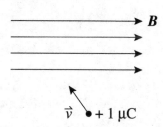

1. A point charge of +1 μC moves with velocity v into a uniform magnetic field B directed to the right, as shown above. What is the direction of the magnetic force on the charge?

   (A) to the right and up the page
   (B) directly out of the page
   (C) directly into the page
   (D) to the right and into the page
   (E) to the right and out of the page

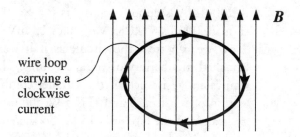

2. A uniform magnetic field B points up the page, as shown above. A loop of wire carrying a clockwise current is placed at rest in this field as shown above, and then let go. Which of the following describes the motion of the wire immediately after it is let go?

   (A) The wire will expand slightly in all directions.
   (B) The wire will contract slightly in all directions.
   (C) The wire will rotate, with the top part coming out of the page.
   (D) The wire will rotate, with the left part coming out of the page.
   (E) The wire will rotate clockwise, remaining in the plane of the page.

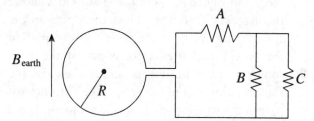

3. An electron moves to the right in a uniform magnetic field that points into the page. What is the direction of the electric field that could be used to cause the electron to travel in a straight line?

   (A) down toward the bottom of the page
   (B) up toward the top of the page
   (C) into the page
   (D) out of the page
   (E) to the left

Free Response:

4. A circular loop of wire of negligible resistance and radius $R = 20$ cm is attached to the circuit shown above. Each resistor has resistance 10 Ω. The magnetic field of the Earth points up along the plane of the page in the direction shown, and has magnitude $B = 5.0 \times 10^{-5}$ T.

   The wire loop rotates about a horizontal diameter, such that after a quarter rotation the loop is no longer in the page, but perpendicular to it. The loop makes 500 revolutions per second, and remains connected to the circuit the entire time.

   (a) Determine the magnetic flux through the loop when the loop is in the orientation shown.
   (b) Determine the maximum magnetic flux through the loop.
   (c) Estimate the average value of the induced EMF in the loop.
   (d) Estimate the average current through resistor C.

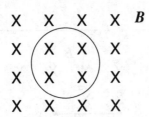

5. A loop of wire is located inside a uniform magnetic field, as shown above. Name at least four things you could do to induce a current in the loop.

› Solutions to Practice Problems

1. C—Use the right-hand rule for the force on charged particles. You point in the direction of the velocity, and curl your fingers in the direction of the magnetic field. This should get your thumb pointing into the page. Because this is a positive charge, no need to switch the direction of the force.

2. C—Use the right-hand rule for the force on a wire. Look at each part of this wire. At the left-most and rightmost points, the current is along the magnetic field lines. Thus, these parts of the wire experience no force. The topmost part of the wire experiences a force out of the page (point to the right, fingers curl up the page, the thumb points out of the page). The bottommost part of the wire experiences a force *into* the page. So, the wire will rotate.

3. A—Use the right-hand rule for the force on a charge. Point in the direction of velocity, curl the fingers into the page, the thumb points up the page . . . but this is a *negative* charge, so the force on the charge is down the page. Now, the electric force must cancel the magnetic force for the charge to move in a straight line, so the electric force should be up the page. (E and B *fields* cannot cancel, but forces sure can.) The direction of an *electric* force on a negative charge is opposite the field; so the field should point down, toward the bottom of the page.

4. (a) Flux equals zero because the field points along the loop, not ever going straight through the loop.

   (b) Flux is maximum when the field *is* pointing straight through the loop; that is, when the loop is perpendicular to the page. Then flux will be just $BA = 5.0 \times 10^{-5} \text{ T} \cdot \pi (0.20 \text{ m})^2 =$

$6.3 \times 10^{-6} \text{ T} \cdot \text{m}^2$. (Be sure your units are right!)

   (c) Induced EMF for this one loop is change in flux over time interval. It takes 1/500 of a second for the loop to make one complete rotation; so it takes $^1/_4$ of that, or 1/2000 of a second, for the loop to go from zero to maximum flux. Divide this change in flux by 1/2000 of a second . . . this is 6.3×10^{-6} $\text{T} \cdot \text{m}^2 / 0.0005 \text{ s} = 0.013 \text{ V}$. (That's 13 mV.)

   (d) Now we can treat the circuit as if it were attached to a battery of voltage 13 mV. The equivalent resistance of the parallel combination of resistors *B* and *C* is 5 Ω; the total resistance of the circuit is 15 Ω. So the current in the whole circuit is $0.013 \text{ V}/15 \text{ W} = 8.4 \times 10^{-4} \text{ A}$. (This can also be stated as 840 μA.) The current splits evenly between resistors *B* and *C* since they're equal resistances, so we get 420 μA for resistor *C*.

5. The question might as well be restated, "name four things you could do to change the flux through the loop," because only a changing magnetic flux induces an EMF.

   (a) Rotate the wire about an axis in the plane of the page. This will change the θ term in the expression for magnetic flux, $BA \cos \theta$.

   (b) Pull the wire out of the field. This will change the area term, because the magnetic field lines will intersect a smaller area of the loop.

   (c) Shrink or expand the loop. This also changes the area term in the equation for magnetic flux.

   (d) Increase or decrease the strength of the magnetic field. This changes the *B* term in the flux equation.

❯ Rapid Review

- Magnetic fields can be drawn as loops going from the north pole of a magnet to the south pole.

- A long, straight, current-carrying wire creates a magnetic field that wraps around the wire in concentric circles. The direction of the magnetic field is found by a right-hand rule.

- Similarly, loops of wire that carry current create magnetic fields. The direction of the magnetic field is, again, found by a right-hand rule.

- A magnetic field exerts a force on a charged particle if that particle is moving perpendicular to the magnetic field.

- When a charged particle moves perpendicular to a magnetic field, it ends up going in circles. This phenomenon is the basis behind mass spectrometry.

- A changing magnetic flux creates an induced EMF, which causes current to flow in a wire.

- Lenz's Law says that when a changing magnetic flux induces a current, the direction of that current will be such that the magnetic field it induces is pointed in the opposite direction of the original change in magnetic flux.

CHAPTER 22

Waves

IN THIS CHAPTER

Summary: A traveling disturbance in a material is a wave. Waves have observable interference properties.

Key Ideas

- ✪ The speed of a wave is constant within a given material. When a wave moves from one material to another, its speed and its wavelength change, but its frequency stays the same.
- ✪ Interference patterns can be observed when a wave goes through two closely spaced slits.
- ✪ Interference patterns can also be observed when light reflects off of a thin film.
- ✪ The Doppler effect causes a moving observer to observe a different frequency than a stationary observer.

Relevant Equations

Speed of a wave:

$$v = \lambda f$$

Position of bright and dark spots for a double slit:

$$x = \frac{m\lambda L}{d} \text{ or } d\sin\theta = m\lambda$$

Index of refraction:

$$n = \frac{c}{v}$$

Wavelength of light in a material:

$$\lambda_n = \frac{\lambda}{n}$$

Thin film interference:

$$2t = m\lambda_n$$

What do a flute, a microwave oven, and a radio all have in common? Well, yes, they all would be nice (if unconventional) birthday presents to receive, but more fundamentally, they all rely on waves. Flutes are designed to produce harmonious sound waves, microwaves use electromagnetic waves to cook food, and radios pick up a different frequency of electromagnetic waves and decode those waves into a Top 40 countdown.

This chapter presents information about what waves are and how they interact. Some of this information will seem familiar—for example, waves are characterized by frequency and amplitude, and these are concepts you already know from simple harmonic motion.

Transverse and Longitudinal Waves

Waves come in two forms: transverse and longitudinal.

> **Wave:** A rhythmic up-and-down or side-to-side motion. Energy is transferred from one particle to another in waves.

A transverse wave occurs when the particles in the wave move perpendicular to the direction of the wave's motion. When you jiggle a string up and down, you create a transverse wave. A transverse wave is shown in Figure 22.1.

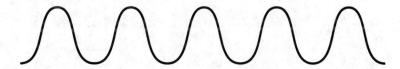

Figure 22.1 Transverse wave.

Longitudinal waves occur when particles move parallel to the direction of the wave's motion. Sound waves are examples of longitudinal waves. A good way to visualize how a sound wave propagates is to imagine one of those "telephones" you might have made when you were younger by connecting two cans with a piece of string.[1] When you talk into one of the cans, your vocal cords cause air molecules to vibrate back and forth, and those vibrating air molecules hit the bottom of the can, which transfers that back-and-forth vibration to the string. Molecules in the string get squished together or pulled apart, depending on whether the bottom of the can is moving back or forth. So the inside of the string would look like Figure 22.2, in which dark areas represent regions where the molecules are squished together, and light areas represent regions where the molecules are pulled apart.

Figure 22.2   Longitudinal wave.

The terms we use to describe waves can be applied to both transverse and longitudinal waves, but they're easiest to illustrate with transverse waves. Take a look at Figure 22.3.

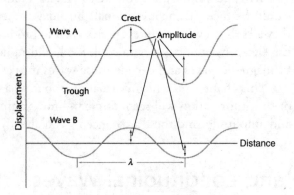

Figure 22.3   Wave terminology.

A **crest** (or a peak) is a high point on a wave, and a **trough** is a low point on a wave. The distance from peak-to-peak or from trough-to-trough is called the **wavelength**. This distance is abbreviated with the Greek letter λ (lambda). The distance that a peak or a trough is from the horizontal axis is called the **amplitude**, abbreviated with the letter A.

The time necessary for one complete wavelength to pass a given point is the period, abbreviated T. The number of wavelengths that pass a given point in one second is the **frequency**, abbreviated f. The period and frequency of a wave are related with a simple equation, which we first saw when we discussed simple harmonic motion:

$$T = \frac{1}{f}$$

If you take a string and jiggle it just once, so that a single peak travels along the string, you generate a **wave pulse**. An example of a wave pulse is shown in Figure 22.4.

---

[1] If you haven't done this before, you should now—it'll make for a fun study break! The easiest way is to take two paper cups, cut a small hole in the bottom of each cup, and connect the cups by running a long piece of string through the holes (be sure to tie a knot at each end of the string so that it can't slide through the holes). Then grab a friend, each of you take a cup, and walk apart until the string is taut. Have your friend whisper into her cup, and if you put your cup up to your ear, you'll hear what she's saying loud and clear . . . or at least loud.

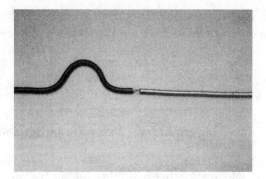

Figure 22.4   Wave pulse.

A **standing wave** occurs when you connect one end of a string to a wall and hold the other end in your hand. As you jiggle the string, waves travel along the string, hit the wall, and then reflect off the wall and travel back to you. If you jiggle the string at just the right frequency, you will notice that the string always looks the same, as in Figure 22.5. This is a standing wave.

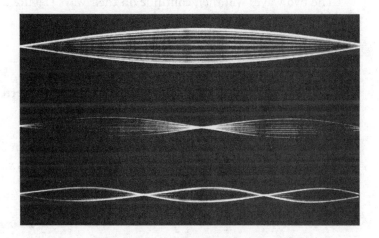

Figure 22.5   Standing wave.

A wave moves a distance λ in a time T. So the velocity at which a wave travels is just distance over time, or λ/T; more simply, this relation is expressed:

$$\boxed{v = \lambda f}$$

For sound waves moving through air at room temperature, $v = 343$ m/s; for light waves in a vacuum, $v = 3.0 \times 10^8$ m/s. Don't memorize these speeds; just be aware.

Interference

When two waves cross each other's path, they interact with each other. There are two ways they can do this—one is called **constructive interference**, and the other is called **destructive interference**.

Constructive interference happens when the peaks of one wave align with the peaks of the other wave. So if we send two wave pulses toward each other as shown in Figure 22.6a,

Figure 22.6a   Two wave pulses about to interfere constructively.

then when they meet in the middle of the string they will interfere constructively (Figure 22.6b).

Figure 22.6b   Two wave pulses interfering constructively.

The two waves will then continue on their ways (Figure 22.6c).

Figure 22.6c   Two wave pulses after interfering constructively.

However, if the peaks of one wave align with the troughs of the other wave, we get destructive interference. For example, if we send the two wave pulses in Figure 22.7a toward each other

Figure 22.7a   Two wave pulses about to interfere destructively.

they will interfere destructively (Figure 22.7b),

Figure 22.7b   Two wave pulses interfering destructively.

and then they will continue along their ways (Figure 22.7c).

Figure 22.7c   Two wave pulses after interfering destructively.

Standing Waves

If you play a stringed instrument, you are already familiar with standing waves. Any time you pluck a guitar string or bow on a cello string, you are setting up a standing wave. Standing waves show up a bit on the AP exam, so they're worth a closer look.

A standing wave occurs when a string is held in place at both ends. When you move the string up and down at certain precise frequencies, you produce a standing wave. Let's first examine several simple standing waves.

Figure 22.8a shows the simplest standing wave, called the fundamental frequency. The wavelength of this wave, λ_1, is twice as long as the distance between the two walls, L.

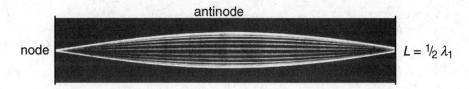

Figure 22.8a   The simplest standing wave.

Notice we labeled our wave with the terms *node* and *antinode*.

Node: Point on a standing wave where the string (or whatever the wave is traveling in) does not move at all

Antinode: Point on a standing wave where the wave oscillates with maximum amplitude

Figure 22.8b shows the next simplest standing wave.

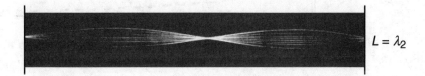

Figure 22.8b   The second simplest standing wave.

Here, the wavelength, λ_2, is exactly equal to the distance between the walls, L. The next standing wave in our progression is drawn in Figure 22.8c.

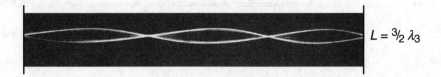

Figure 22.8c   The next standing wave in our progression.

Okay, we have a pattern developing here. Notice that the relationship between L, the distance between the walls, and the wavelength of the standing wave is

$$L = \frac{n}{2}\lambda_n$$

We can manipulate this equation to say

$$\lambda_n = \frac{2L}{n}$$

And we know that the velocity of a wave is $v = \lambda f$. So we can do some more manipulation of variables and come up with this equation:

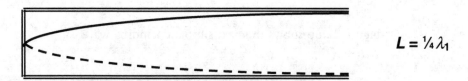

$$f_n = \frac{nv}{2L} \qquad \Longleftarrow \boxed{\text{String or Open Pipe}}$$

This equation says that the frequency of the 1st, 2nd, 3rd, or mth standing wave equals n, multiplied by the velocity of the wave, v, divided by two times the distance between the walls, L.

There are two uses of this equation. One is for stringed instruments—the illustrations on the previous pages could be viewed as a guitar string being plucked—and one is for sound in a pipe open at both ends.

If we have sound waves in a pipe that's closed at one end, the situation looks slightly different. The fundamental frequency is shown in Figure 22.9a.

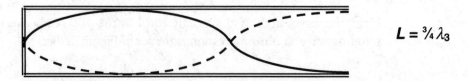

$$L = \frac{1}{4}\lambda_1$$

Figure 22.9a   Fundamental frequency in a closed pipe.

And the next simplest standing wave is shown in Figure 22.9b.

$$L = \frac{3}{4}\lambda_3$$

Figure 22.9b   Next simplest standing wave in a closed pipe.

Notice that we did not have a λ_2. When one end of a pipe is closed, we can only have odd values for n. The frequency of the mth standing wave in a closed pipe is

$$f_n = \frac{nv}{4L}, \text{ where } n \text{ must be } odd.$$

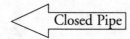 Closed Pipe

When you have two sound waves of almost, but not quite, equal frequency, you may hear **beats**.

Beats: Rhythmic interference that occurs when two notes of unequal but close frequencies are played

---

If you have a couple of tuning forks of similar—but not identical—frequency to play with, or if you have a couple of tone generators at your disposal, you might enjoy generating some beats of your own. They make a wonderful "wa-wa" sound, which is due to a periodic increase and decrease in intensity, or loudness. The frequency of the "wa-wa" is equal to the *difference* between the two frequency generators.

Doppler Effect

Whenever a fire engine or ambulance races by you with its sirens blaring, you experience the **Doppler effect**. Similarly, if you enjoy watching auto racing, that "Neeee-yeeeer" you hear as the cars scream past the TV camera is also attributable to the Doppler effect.

Doppler Effect: The apparent change in a wave's frequency that you observe whenever the source of the wave is moving toward or away from you

---

To understand the Doppler effect, let's look at what happens as a fire truck travels past you (Figures 22.10a and 22.10b).

As the fire truck moves toward you, its sirens are emitting sound waves. Let's say that the sirens emit one wave pulse when the truck is 50 meters away from you. It then emits another pulse when the truck is 49.99 meters away from you. And so on. Because the truck keeps moving toward you as it emits sound waves, it appears to you that these waves are getting scrunched together.

Figure 22.10a   When the fire truck moves toward you, the sound waves get squished together, increasing the frequency you hear.

Then, once the truck passes you and begins to move away from you, it appears as if the waves are stretched out.

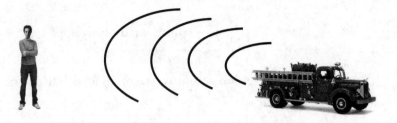

Figure 22.10b   As the fire truck moves away from you, the sound waves spread apart, and you hear a lower frequency.

Now, imagine that you could record the instant that each sound wave hit you. When the truck was moving toward you, you would observe that the time between when one wave hit and when the next wave hit was very small. However, when the truck was moving away from you, you would have to wait a while between when one wave hit you and when the next one did. In other words, when the truck was moving toward you, you registered that the sirens were making a higher frequency noise; and when the truck was moving away, you registered that the sirens were making a lower frequency noise.

That's all you really need to know about the Doppler effect. Just remember, the effect is rather small—at normal speeds, the frequency of, say, a 200 Hz note will only change by a few tens of Hz, not hundreds of Hz.

Electromagnetic Waves

When radio waves are beamed through space, or when X-rays are used to look at your bones, or when visible light travels from a light bulb to your eye, electromagnetic waves are at work. All these types of radiation fall in the electromagnetic spectrum, shown in Figure 22.11.

AM Radio	FM Radio and TV	Microwaves	Infrared	Visible Light	Ultraviolet	X-rays	Gamma Rays

← ——————————————— Increasing frequency ——————————————— →

Figure 22.11   The electromagnetic spectrum.

The unique characteristic about electromagnetic waves is that all of them travel at exactly the same speed through a vacuum—3×10^8 m/s. The more famous name for "3×10^8 m/s" is "the speed of light," or "c."

What makes one form of electromagnetic radiation different from another form is simply the frequency of the wave. AM radio waves have a very low frequency and a very long wavelength; whereas, gamma rays have an extremely high frequency and an exceptionally short wavelength. But they're all just varying forms of light waves.

Single and Double Slits

Long ago, when people still used words like "thence" and "hither," scientists thought that light was made of particles. About a hundred years ago, at the beginning of the twentieth century, however, scientists began to change their minds. "Light is a wave!" they proclaimed.

Then such physicists as Einstein said, "Actually, light acts as either a particle or a wave, depending on what methods you're using to detect it."[2] But for now, let's stick to the turn-of-the-twentieth-century notion that light is simply a wave.

The way that physicists showed that light behaves like a wave was through slit experiments. Consider light shining through two very small slits, located very close together—slits separated by tenths or hundredths of millimeters. The light shone through each slit and then hit a screen. But here's the kicker: rather than seeing two bright patches on the screen (which would be expected if light was made of particles), the physicists saw lots of bright patches. The only way to explain this phenomenon was to conclude that light behaves like a wave.

Look at Figure 22.12a. When the light waves went through each slit, they were diffracted. As a result, the waves that came through the top slit interfered with the waves that came through the bottom slit—everywhere that peaks or troughs crossed paths, either constructive or destructive interference occurred.

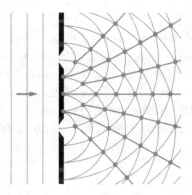

Figure 22.12a   Light waves interacting in the double-slit experiment.

So when the light waves hit the screen, at some places they constructively interfered with one another and in other places they destructively interfered with one another. That explains why the screen looked like Figure 22.12b.

Figure 22.12b   Image on the screen in the double-slit experiment.

---

[2] Don't worry about this now. But it's a very cool concept, so if you're interested, ask your physics teacher to recommend a book on twentieth century physics. There are several very approachable books available on this subject. (We've listed a few of our favorite books on modern physics in the Bibliography.)

The bright areas were where constructive interference occurred, and the dark areas were where destructive interference occurred. Particles can't interfere with one another—only waves can—so this experiment proved that light behaves like a wave.

When light passes through slits to reach a screen, the equation to find the location of bright spots is as follows.

$$d \sin \theta = m\lambda$$

Here, d is the distance between slits, λ is the wavelength of the light, and m is the "order" of the bright spot; we discuss m below. θ is the angle at which an observer has to look to see the bright spot. Usually, the bright spots are pretty close together, and almost directly across from the slits. In many cases, the angle θ is small, so we can use the following equation instead. It describe where on the screen you would find patches of constructive or destructive interference.

$$x = \frac{m\lambda L}{d}$$

In these equations, x is the distance *from the center of the screen* wherein you would find the region you're seeking (either a bright region or a dark region), λ is the wavelength of the light shining through the slits, L is the distance between the slits and the screen, and d is the distance between the slits. The variable m represents the "order" of the bright or dark spot, measured from the central maximum as shown in Figure 22.13. Bright spots get integer values of m; dark spots get half-integer values of m. The central maximum represents $m = 0$.

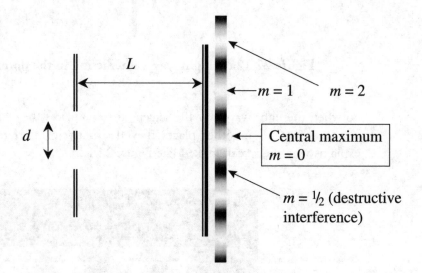

Figure 22.13　Variables used in the double-slit equations.

So, for example, if you wanted to find how far from the center of the pattern the first bright spot labeled $m = 1$ is, you would plug in "1" for m, If you wanted to find the dark region closest to the center of the screen, you would plug in "½" for m.

Single Slits and Diffraction Gratings

Once you understand the double-slit experiment, single slits and diffraction gratings are simple.

A diffraction grating consists of a large number of slits, not just two slits. The locations of bright and dark spots on the screen are the same as for a double slit, but the bright spots produced by a diffraction grating are very sharp dots.

A single slit produces interference patterns, too, because the light that bends around each side of the slit interferes upon hitting the screen. For a single slit, the central maximum is bright and very wide; the other bright spots are regularly spaced, but dim relative to the central maximum.

Index of Refraction

Light also undergoes interference when it reflects off of thin films of transparent material. Before studying this effect quantitatively, though, we have to examine how light behaves when it passes through different materials.

Light—or any electromagnetic wave—travels at the speed c, or 3×10^8 m/s. But it only travels at this speed through a vacuum, when there aren't any pesky molecules to get in the way. When it travels through anything other than a vacuum, light slows down. The amount by which light slows down in a material is called the material's **index of refraction**.

> **Index of refraction:** A number that describes by how much light slows down when it passes through a certain material, abbreviated n

The index of refraction can be calculated using this equation.

$$n = \frac{c}{v}$$

This says that the index of refraction of a certain material, n, equals the speed of light in a vacuum, c, divided by the speed of light through that material, v.

For example, the index of refraction of glass is about 1.5. This means that light travels 1.5 times faster through a vacuum than it does through glass. The index of refraction of air is approximately 1. Light travels through air at just about the same speed as it travels through a vacuum.

TIP Another thing that happens to light as it passes through a material is that its wavelength changes. When light waves go from a medium with a low index of refraction to one with a high index of refraction, they get squished together. So, if light waves with a wavelength of 500 nm travel through air ($n_{air} = 1$), enter water ($n_{water} = 1.33$), and then emerge back into air again, it would look like Figure 22.14.

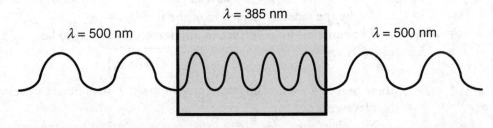

$\lambda = 385$ nm

$\lambda = 500$ nm $\lambda = 500$ nm

Figure 22.14   Light passing through air, into water, and then back into air again. The wavelength of the light changes as it goes from one medium to another.

The equation that goes along with this situation is the following.

$$\lambda_n = \frac{\lambda}{n}$$

In this equation, λ_n is the wavelength of the light traveling through the transparent medium (like water, in Figure 22.14), λ is the wavelength in a vacuum, and n is the index of refraction of the transparent medium.

It is important to note that, even though the wavelength of light changes as it goes from one material to another, its frequency remains constant. The frequency of light is a property of the photons that comprise it (more about that in Chapter 24), and the frequency doesn't change when light slows down or speeds up.

Thin Films

When light hits a thin film of some sort, the interference properties of the light waves are readily apparent. You have likely seen this effect if you've ever noticed a puddle in a parking lot. If a bit of oil happens to drop on the puddle, the oil forms a very thin film on top of the water. White light (say, from the sun) reflecting off of the oil undergoes interference, and you see some of the component colors of the light.

Consider a situation where monochromatic light (meaning "light that is all of the same wavelength") hits a thin film, as shown in Figure 22.15. At the top surface, some light will be reflected, and some will penetrate the film. The same thing will happen at the bottom surface: some light will be reflected back up through the film, and some will keep on traveling out through the bottom surface. Notice that the two reflected light waves overlap; the wave that reflected off the top surface and the wave that traveled through the film and reflected off the bottom surface will interfere.

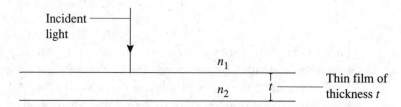

Figure 22.15   Monochromatic light hitting a thin film.

The important thing to know here is whether the interference is constructive or destructive. The wave that goes through the film travels a distance of $2t$ before interfering, where t is the thickness of the film. If this extra distance is precisely equal to a wavelength, then the interference is constructive. You also get constructive interference if this extra distance is precisely equal to two, or three, or any whole number of wavelengths.

But be careful what wavelength you use . . . because this extra distance occurs inside the film, we're talking about the wavelength *in the film*, which is the wavelength in a vacuum divided by the index of refraction of the film.

The equation for constructive interference turns out to be

$$2t = m\lambda_n$$

where m is any whole number, representing how many extra wavelengths the light inside the film went.

So, when does *destructive* interference occur? When the extra distance in the film precisely equals a half wavelength . . . or one and a half wavelengths, or two and a half wavelengths . . . so for destructive interferences, plug in a half-integer for m.

There's one more complication. If light reflects off of a surface while going from low to high index of refraction, the light "changes phase." For example, if light in air reflects off oil ($n \sim 1.2$), the light changes phase. If light in water reflects off oil, though, the light does not change phase. For our purposes, a phase change simply means that the conditions for constructive and destructive interference are reversed.

Summary: For thin film problems, go through these steps.

1. Count the phase changes. A phase change occurs for every reflection from low to high index of refraction.
2. The extra distance traveled by the wave in the film is twice the thickness of the film.
3. The wavelength in the film is $\lambda_n = \lambda/n$, where n is the index of refraction of the film's material.
4. Use the equation $2t = m\lambda_n$. If the light undergoes zero or two phase changes, then plugging in whole numbers for m gives conditions for constructive interference. (Plugging in half-integers for m gives destructive interference.) If the light undergoes one phase change, conditions are reversed—whole numbers give destructive interference, half-integers, constructive.

Finally, why do you see a rainbow when there's oil on top of a puddle? White light from the sun, consisting of *all* wavelengths, hits the oil. The thickness of the oil at each point only allows one wavelength to interfere constructively. So, at one point on the puddle, you just see a certain shade of red. At another point, you just see a certain shade of orange, and so on. The result, over the area of the entire puddle, is a brilliant, swirling rainbow.

› Practice Problems

Multiple Choice:

1. A talk show host inhales helium; as a result, the pitch of his voice rises. What happens to the standing waves in his vocal cords to cause this effect?

   (A) The wavelength of these waves increases.
   (B) The wavelength of these waves decreases.
   (C) The speed of these waves increases.
   (D) The speed of these waves decreases.
   (E) The frequency of these waves decreases.

2. In a pipe closed at one end and filled with air, a 384-Hz tuning fork resonates when the pipe is 22-cm long; this tuning fork does not resonate for any smaller pipes. For which of these closed pipe lengths will this tuning fork also resonate?

   (A) 11 cm
   (B) 44 cm
   (C) 66 cm
   (D) 88 cm
   (E) 384 cm

3. Monochromatic light passed through a double slit produces an interference pattern on a screen a distance 2.0 m away. The third-order maximum is located 1.5 cm away from the central maximum. Which of the following adjustments would cause the third-order maximum instead to be located 3.0 cm from the central maximum?

   (A) doubling the distance between slits
   (B) tripling the wavelength of the light
   (C) using a screen 1.0 m away from the slits
   (D) using a screen 3.0 m away from the slits
   (E) halving the distance between slits

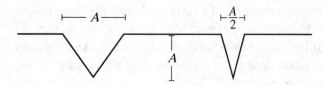

4. The two wave pulses shown above are moving toward each other along a string. When the two pulses interfere, what is the maximum amplitude of the resultant pulse?

(A) $(1/2)A$
(B) A
(C) $(3/2)A$
(D) $2A$
(E) $(5/2)A$

Free Response:

5. Laser light is passed through a diffraction grating with 7000 lines per centimeter. Light is projected onto a screen far away. An observer by the diffraction grating observes the first order maximum 25° away from the central maximum.

(a) What is the wavelength of the laser?
(b) If the first order maximum is 40 cm away from the central maximum on the screen, how far away is the screen from the diffraction grating?
(c) How far, measured along the screen, from the central maximum will the second-order maximum be?

› Solutions to Practice Problems

1. C—The frequency of these waves must go up, because the pitch of a sound is determined by its frequency. The wavelength of the waves in the host's voice box doesn't change, though, because the wavelength is dependent on the physical structure of the host's body. Thus, by $v = \lambda f$, the speed of the waves in his vocal cords must go up. You can even look it up—the speed of sound in helium is faster than the speed of sound in normal air.

2. C—A wave in a pipe closed at one end has a node at one end and an antinode at the other. 22 cm is the length of the pipe for the fundamental oscillation, which looks like this:

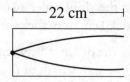

You can see that $1/2$ of a "hump" is contained in the pipe, so the total wavelength (two full "humps") must be 88 cm. The next harmonic oscillation occurs when there is again a node at one end of the pipe and an antinode at the other, like this:

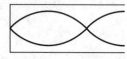

This pipe contains $1^1/2$ "humps," so its length equals three-quarters of the total wavelength. The pipe length is thus 66 cm.

3. E—Use the equation

$$x = \frac{m\lambda L}{d}$$

Here $m = 3$ because we are dealing with the third-order maximum. We want to double the distance to this third-order maximum, which means we want to double x in the equation. To do this, halve the denominator; d in the denominator represents the distance between slits.

4. D—When wave pulses interfere, their amplitudes add algebraically. The question asks for the maximum amplitude, so the widths of the pulses are irrelevant. When both pulses are *right* on top of one another, each pulse will have amplitude A; these amplitudes will add to a resultant of $2A$.

5. (a) Use $d \sin \theta = m\lambda$. Here d is *not* 7000! d represents the distance between slits. Because there are 7000 lines per centimeter, there's 1/7000 centimeter per line; thus, the distance between lines is 1.4×10^{-4} cm, or 1.4×10^{-6} m. θ is 25° for the first-order maximum, where $m = 1$. Plugging in, you get a wavelength of just about 6×10^{-7} m, also known as 600 nm.

(b) This is a geometry problem.

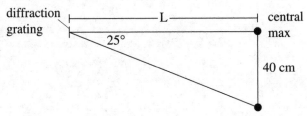

$\tan 25° = (40 \text{ cm})/L$; solve for L to get 86 cm, or about 3 feet.

(c) Use $d \sin \theta = m\lambda$; solve for θ using $m = 2$, and convert everything to meters. We get $\sin \theta = 2(6.0 \times 10^{-7} \text{ m})/(1.4 \times 10^{-6} \text{ m})$. The angle will be 59°. Now, use the same geometry from part (b) to find the distance along the screen: $\tan 59° = x/(0.86 \text{ m})$, so $x = 143$ cm. (Your answer will be counted correct if you rounded differently and just came close to 143 cm.)

› Rapid Review

- Waves can be either transverse or longitudinal. Transverse waves are up–down waves, like a sine curve. Longitudinal waves are push–pull waves, like a sound wave traveling through the air.

- When two waves cross paths, they can interfere either constructively or destructively. Constructive interference means that the peaks of the waves line up, so when the waves come together, they combine to make a wave with bigger amplitude than either individual wave. Destructive interference means that the peak of one wave lines up with the trough of the other, so when the waves come together, they cancel each other out.

- Standing waves occur when a wave in a confined space (for example, on a violin string or in a pipe) has just the right frequency so that it looks the same at all times. The simplest standing wave possible in a situation is called the fundamental frequency.

- The Doppler effect describes what happens when the source of a wave—such as a fire truck's siren—is moving relative to you. If the source moves toward you, you perceive the waves to have a higher frequency than they really do, and if the source moves away from you, you perceive the waves to have a lower frequency than they really do.

- All electromagnetic waves travel at a speed of 3×10^8 m/s in a vacuum.

- The double-slit experiment demonstrates that light behaves like a wave.

- When light travels through anything other than a vacuum, it slows down, and its wavelength decreases. The amount by which light slows down as it passes through a medium (such as air or water) is related to that medium's index of refraction.

- Thin films can cause constructive or destructive interference, depending on the thickness of the film and the wavelength of the light. When solving problems with thin films, remember to watch out for phase changes.

CHAPTER **23**

Optics

IN THIS CHAPTER

Summary: Light rays travel in a straight line, except when they're bent or reflected. The rules of refraction and reflection can be used to figure out how the light's path changes.

Key Ideas

- ✪ When light hits an interface between materials, the light reflects and refracts. The angle of refraction is given by Snell's law.
- ✪ Total internal reflection can occur when light strikes a boundary going from a higher-index material to a lower-index material at a large incident angle.
- ✪ Concave mirrors and convex lenses are optical instruments that converge light to a focal point. Concave lenses and convex mirrors are optical instruments that cause light to diverge.
- ✪ Virtual images are formed right-side up; real images are upside down and can be projected on a screen.

Relevant Equations

Snell's law (for refraction):

$$n_1 \sin\theta_1 = n_2 \sin\theta_2$$

Finding the critical angle for total internal reflection:

$$\sin\theta_c = \frac{n_2}{n_1}$$

The lensmaker's equation/mirror equation:

$$\frac{1}{f} = \frac{1}{d_i} + \frac{1}{d_o}$$

Magnification equation:

$$m = -\frac{d_i}{d_o} = \frac{h_i}{h_o}$$

Optics is the study of light and how it gets bent and reflected by lenses and mirrors. If you're tired of problems involving heavy calculation, optics will come as a breath of fresh air. It will feel like a springtime breeze, a splash of cool water, an Enya song. Why? Because, for the most part, optics problems will not require a lot of math. Instead, you'll have to draw pictures. (Finally, a physics topic for the artist-at-heart!)

Snell's Law

In addition to changing its speed and its wavelength, light can also change its direction when it travels from one medium to another. The way in which light changes its direction is described by Snell's law.

$$n_1 \sin\theta_1 = n_2 \sin\theta_2$$

To understand Snell's law, it's easiest to see it in action. Figure 23.1 should help.

In Figure 23.1, a ray of light is going from air into water. The dotted line perpendicular to the surface is called the "normal."[1] This line is not real; rather it is a reference line for use in Snell's law. *In optics, ALL ANGLES ARE MEASURED FROM THE NORMAL, NOT FROM A SURFACE!*

As the light ray enters the water, it is being bent toward the normal. The angles θ_1 and θ_2 are marked on the figure, and the index of refraction of each material, n_1 and n_2, is also noted. If we knew that θ_1 equals 55°, for example, we could solve for θ_2 using Snell's law.

$$(1.00)(\sin 55°) = (1.33)(\sin \theta_2)$$

$$\theta_2 = 38°$$

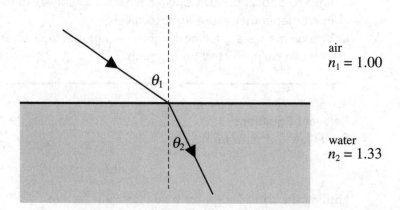

Figure 23.1   Light bending as it travels from one material (air) to another (water).

---

[1] Remember that in physics, "normal" means "perpendicular"—the normal is always perpendicular to a surface.

Whenever light goes from a medium with a low index of refraction to one with a high index of refraction—as in our drawing—the ray is bent toward the normal. Whenever light goes in the opposite direction—say, from water into air—the ray is bent away from the normal.

If you have a laser pointer, try shining it into some slightly milky water . . . you'll see the beam bend into the water. But you'll also see a little bit of the light reflect off the surface, at an angle equal to the initial angle. (Careful the reflected light doesn't get into your eye!) In fact, at a surface, if light is refracted into the second material, some light must be reflected.

Sometimes, though, when light goes from a medium with a high index of refraction to one with a low index of refraction, we could get **total internal reflection**. For total internal reflection to occur, the light ray must be directed at or beyond the **critical angle**.

> **Critical Angle:** The angle past which rays cannot be transmitted from one material to another, abbreviated θ_c

$$\sin \theta_c = \frac{n_2}{n_1}$$

Again, pictures help, so let's take a look at Figure 23.2.

In Figure 23.2, a ray of light shines up through a glass block. The critical angle for light going from glass to air is 42°; however, the angle of the incident ray is greater than the critical angle. Therefore, the light cannot be transmitted into the air. Instead, all of it reflects inside the glass. Total internal reflection occurs anytime light cannot leave a material.

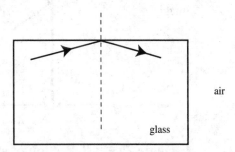

Figure 23.2   Light undergoing total internal reflection.

Mirrors

Okay, time to draw some pictures. Let's start with plane mirrors.

The key to solving problems that involve plane mirrors is that the angle at which a ray of light hits the mirror equals the angle at which it bounces off, as shown in Figure 23.3.

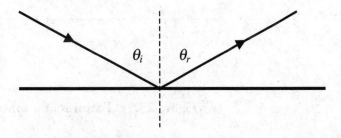

Figure 23.3   Light reflecting off a plane mirror.

In other words—or, more accurately, "in other symbols"—$\theta_i = \theta_r$, where θ_i is the incident[2] angle, and θ_r is the reflected angle.

So, let's say you had an arrow, and you wanted to look at its reflection in a plane mirror. We'll draw what you would see in Figure 23.4.

The image of the arrow that you would see is drawn in Figure 23.4 in dotted lines. To draw this image, we first drew the rays of light that reflect from the top and bottom of the arrow to your eye. Then we extended the reflected rays through the mirror.

Whenever you are working with a plane mirror, follow these rules:

- The image is upright. Another term for an upright image is a **virtual image**.
- The image is the same size as the original object. That is, the magnification, m, is equal to 1.
- The image distance, d_i, equals the object distance, d_o.

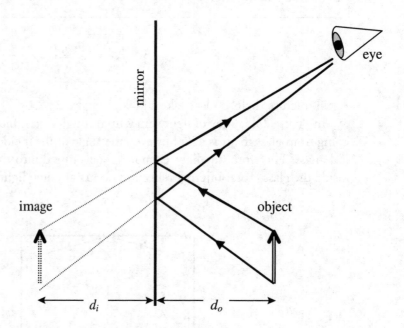

Figure 23.4   Reflection of an arrow in a plane mirror.

A more challenging type of mirror to work with is called a spherical mirror. Before we draw our arrow as it looks when reflected in a spherical mirror, let's first review some terminology (this terminology is illustrated in Figure 23.5).

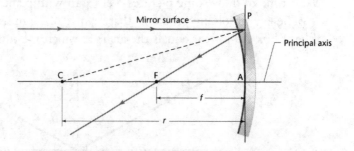

Figure 23.5   Features of a spherical mirror.

---

[2] The "incident" angle simply means "the angle that the initial ray happened to make."

A spherical mirror is a curved mirror—like a spoon—that has a constant radius of curvature, *r*. The imaginary line running through the middle of the mirror is called the **principal axis**. The point labeled "C," which is the center of the sphere, lies on the principal axis and is located a distance *r* from the middle of the mirror. The point labeled "F" is the focal point, and it is located a distance *f*, where $f = (r/2)$, from the middle of the mirror. The focal point is also on the principal axis. The line labeled "P" is perpendicular to the principal axis.

There are several rules to follow when working with spherical mirrors. Memorize these.

- Incident rays that are parallel to the principal axis reflect through the focal point.
- Incident rays that go through the focal point reflect parallel to the principal axis.
- Any points that lie on the same side of the mirror as the object are a *positive* distance from the mirror. Any points that lie on the other side of the mirror are a *negative* distance from the mirror.
- $\dfrac{1}{f} = \dfrac{1}{d_o} + \dfrac{1}{d_i}$.

That last rule is called the "mirror equation." (You'll find this equation to be identical to the "lensmaker's equation" later.)

To demonstrate these rules, we'll draw three different ways to position our arrow with respect to the mirror. In the first scenario, we'll place our arrow on the principal axis, beyond point "C," as shown in Figure 23.6.

Notice that the image here is upside down. Whenever an image is upside down, it is called a **real image**. A real image can be projected onto a screen, whereas a virtual image cannot.[3]

The magnification, *m*, is found by

$$m = -\frac{d_i}{d_o}$$

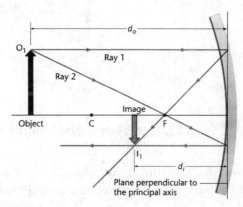

Figure 23.6 Object located beyond the center point of a spherical mirror.

---

[3] TRY THIS! Light a candle; reflect the candle in a concave mirror; and put a piece of paper where the image forms. . . . You will actually see a picture of the upside-down, flickering candle!

When the magnification is negative, the image is upside down. By convention, the height of an upside-down image is a negative value. When the magnification is positive, the image is right-side up, and the height is a positive value. Plugging in values from our drawing in Figure 23.6, we see that our magnification should be less than 1, and the image should be upside down. Which it is. Good for us.

Now we'll place our arrow between the mirror and "F," as shown in Figure 23.7. When an object is placed between "F" and the mirror, the image created is a virtual image—it is upright.

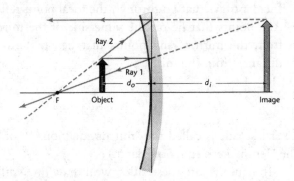

**Figure 23.7   Object located between focal point
of a spherical mirror and the mirror itself.**

Finally, we will place our object on the other side of the mirror, as shown in Figure 23.8. Now, in Figure 23.8, we have a convex mirror, which is a *diverging* mirror—parallel rays tend to spread away from the focal point. In this situation, the image is again virtual.

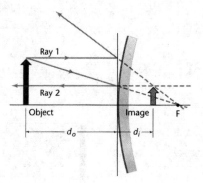

Figure 23.8   Object located on the convex side of a spherical mirror.

When we use the mirror equation here, we have to be especially careful: d_o is positive, but both d_i and f are negative, because they are not on the same side of the mirror as the object.

Note that the convex (diverging) mirror *cannot* produce a real image. Give it a try—you can't do it!

Lenses

We have two types of lenses to play with: convex and concave. A **convex lens**, also known as a "converging lens," is shown in Figure 23.9a.

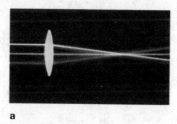

Figure 23.9a   Convex lens.

And a **concave lens**, or "diverging lens," is shown in Figure 23.9b.

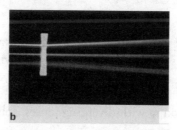

Figure 23.9b   Concave lens.

We'll start by working with a convex lens. The rules to follow with convex lenses are these:

- An incident ray that is parallel to the principal axis refracts through the far focal point.
- An incident ray that goes through near focal point refracts parallel to the principal axis.
- The lensmaker's equation and the equation to find magnification are the same as for mirrors. In the lensmaker's equation, f is positive.

Want to try these rules out? Sure you do. We'll start, in Figure 23.10, by placing our arrow farther from the lens than the focal point.

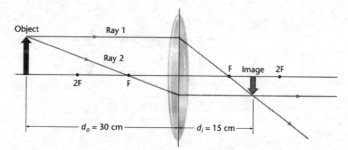

Figure 23.10   Object located farther from a convex lens than the focal point.

We could also demonstrate what would happen if we placed our object in between the near focal point and the lens. But so could you, as long as you follow our rules. And we don't want to stifle your artistic expression. So go for it.[4]

---

[4] Answer: The image is a virtual image, located on the same side of the lens as the object, but farther from the lens than the object.

Now, how about a problem?

> A 3-cm-tall object is placed 20 cm from a converging lens. The focal distance of the lens is 10 cm. How tall will the image be?

We are given d_o and f. So we have enough information to solve for d_i using the lensmaker's equation.

$$\frac{1}{0.1} = \frac{1}{0.2} + \frac{1}{d_i}$$

Solving, we have $d_i = 20$ cm. Now we can use the magnification equation.

$$m = -\frac{0.2}{0.2} = -1$$

Our answer tells us that the image is exactly the same size as the object, but the negative sign tells us that the image is upside down. So our answer is that the image is 3 cm tall, and that it is real.

When working with diverging lenses, follow these rules:

- An incident ray parallel to the principal axis will refract as if it came from the near focal point.
- An incident ray toward the far focal point refracts parallel to the principal axis.
- The lensmaker's equation and the magnification equation still hold true. With diverging lenses, though, f is negative.

We'll illustrate these rules by showing what happens when an object is placed farther from a concave lens than the focal point. This is shown in Figure 23.11.

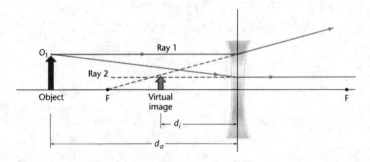

Figure 23.11   Object located farther from a concave lens than the focal point.

The image in Figure 23.11 is upright, so we know that it is virtual.

Now go off and play with lenses. And spoons. And take out that box of crayons that has been collecting dust in your cupboard and draw a picture. Let your inner artist go wild. (Oh, and do the practice problems, too!)

› Practice Problems

Multiple Choice:

1. In an aquarium, light traveling through water ($n = 1.3$) is incident upon the glass container ($n = 1.5$) at an angle of 36° from the normal. What is the angle of transmission in the glass?

   (A) The light will not enter the glass because of total internal reflection.
   (B) 31°
   (C) 36°
   (D) 41°
   (E) 52°

2. Which of the following optical instruments can produce a virtual image with magnification 0.5?

   I.   convex mirror
   II. concave mirror
   III. convex lens
   IV. concave lens

   (A) I and IV only
   (B) II and IV only
   (C) I and II only
   (D) III and IV only
   (E) I, II, III, and IV

3. Light waves traveling through air strike the surface of water at an angle. Which of the following statements about the light's wave properties upon entering the water is correct?

   (A) The light's speed, frequency, and wavelength all stay the same.
   (B) The light's speed, frequency, and wavelength all change.
   (C) The light's speed and frequency change, but the wavelength stays the same.
   (D) The light's wavelength and frequency change, but the light's speed stays the same.
   (E) The light's wavelength and speed change, but the frequency stays the same.

4. An object is placed at the center of a concave spherical mirror. What kind of image is formed, and where is that image located?

   (A) A real image is formed at the focal point of the mirror.
   (B) A real image is formed at the center of the mirror.
   (C) A real image is formed one focal length beyond the center of the mirror.
   (D) A virtual image is formed one focal length behind the mirror.
   (E) A virtual image is formed one radius behind the mirror.

Free Response:

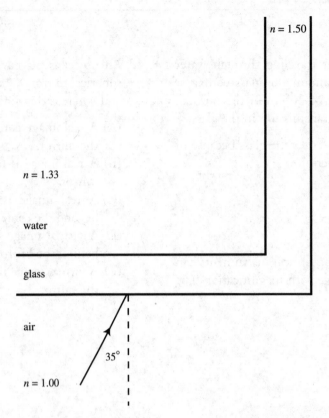

5. Light traveling through air encounters a glass aquarium filled with water. The light is incident on the glass from the front at an angle of 35°.

(a) At what angle does the light enter the glass?
(b) At what angle does the light enter the water?
(c) On the diagram above, sketch the path of the light as it travels from air to water. Include all reflected and refracted rays; label all angles of reflection and refraction.

After entering the water, the light encounters the side of the aquarium, hence traveling back from water to glass. The side of the tank is perpendicular to the front.

(d) At what angle does light enter the glass on the side of the aquarium?
(e) Does the light travel out of the glass and into the air, or does total internal reflection occur? Justify your answer.

› Solutions to Practice Problems

1. **B**—If you had a calculator, you could use Snell's law, calling the water medium "1" and the glass medium "2": 1.3 sin 36° = 1.5 sin θ_2. You would find that the angle of transmission is 31°. But, you don't have a calculator . . . so look at the choices. The light must bend *toward* the normal when traveling into a material with higher index of refraction, and choice B is the only angle smaller than the angle of incidence. Choice A is silly because total internal reflection can only occur when light goes from high to low index of refraction.

2. **A**—The converging optical instruments—convex lens and concave mirror—only produce virtual images if the object is inside the focal point. But when that happens, the virtual image is *larger* than the object, as when you look at yourself in a spoon or a shaving mirror. But a diverging optical instrument—a convex mirror and a concave lens—always produces a smaller, upright image, as when you look at yourself reflected in a Christmas tree ornament.

3. **E**—The speed of light (or any wave) depends upon the material through which the wave travels; by moving into the water, the light's speed slows down. But the frequency of a wave does *not* change, even when the wave changes material. This is why tree leaves still look green under water—color is determined by frequency, and the frequency of light under water is the same as in air. So, if speed changes and frequency stays the same, by $v = \lambda f$, the wavelength must also change.

4. **B**—You could approximate the answer by making a ray diagram, but the mirror equation works, too:

$$\frac{1}{f} = \frac{1}{d_o} + \frac{1}{d_i}$$

Because the radius of a spherical mirror is twice the focal length, and we have placed the object at the center, the object distance is equal to $2f$. Solve the mirror equation for d_i by finding a common denominator:

$$d_i = \frac{fd_o}{d_o - f}$$

This works out to $(f)(2f)/(2f - f)$ which is just $2f$. The image distance is twice the focal length, and at the center point. This is a real image because d_i is positive.

5. (a) Use Snell's law: $n_1 \sin \theta_1 = n_2 \sin \theta_2$. This becomes 1.0 sin 35° = 1.5 sin θ_2. Solve for θ_2 to get 22°.

(b) Use Snell's law again. This time, the angle of incidence on the water is equal to the angle of refraction in the glass, or 22°. The angle of refraction in water is 25°. This makes sense because light should bend away from normal when entering the water because water has smaller index of refraction than glass.

(c) Important points:
*Light both refracts *and* reflects at both surfaces. You must show the reflection, with the angle of incidence equal to the angle of reflection.
*We know you don't have a protractor, so the angles don't have to be perfect. But the light must bend toward normal when entering glass, and away from normal when entering water.

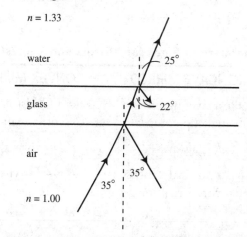

(d) The angle of incidence on the side must be measured from the normal. The angle of incidence is not 25°, then, but 90 − 25 = 65°. Using Snell's law, 1.33 sin 65° = 1.50 sin θ_2. The angle of refraction is 53°.

(e) The critical angle for glass to air is given by sin θ_c = 1.0/1.5. So θ_c = 42°. Because the angle of incidence from the glass is 53° [calculated in (d)], total internal reflection occurs.

❯ Rapid Review

- Snell's law describes how the direction of a light beam changes as it goes from a material with one index of refraction to a material with a different index of refraction.

- When light is directed at or beyond the critical angle, it cannot pass from a material with a high index of refraction to one with a low index of refraction; instead, it undergoes total internal reflection.

- When solving a problem involving a plane mirror, remember (1) the image is upright (it is a virtual image); (2) the magnification equals 1; and (3) the image distance equals the object distance.

- When solving a problem involving a spherical mirror, remember (1) incident rays parallel to the principal axis reflect through the focal point; (2) incident rays going through the focal point reflect parallel to the principal axis; (3) points on the same side of the mirror as the object are a positive distance from the mirror, and points on the other side are a negative distance from the mirror; and (4) the lensmaker's equation (also called the mirror equation in this case) holds.

- When solving problems involving a convex lens, remember (1) incident rays parallel to the principal axis refract through the far focal point; (2) incident rays going through the near focal point refract parallel to the principal axis; and (3) the lensmaker's equation holds, and f is positive.

- When solving problems involving a concave lens, remember (1) incident rays parallel to the principal axis refract as if they came from the near focal point; (2) incident rays going toward the far focal point refract parallel to the principal axis; and (3) the lensmaker's equation holds with a negative f.

Summary of Signs for Mirrors and Lenses

CONCAVE MIRROR	CONVEX MIRROR	CONCAVE LENS	CONVEX LENS
Like a shaving mirror or makeup compact	Like a Christmas tree ornament	Like a pair of glasses	Like a magnifying glass
d_o is always +	d_o is always +	d_o is always +	d_o is always +
d_i is + for real images and − for virtual images	d_i is − for virtual images	d_i is − for virtual images	d_i is + for real images and − for virtual images
f is always +	f is always −	f is always −	f is always +

Atomic and Nuclear Physics

IN THIS CHAPTER

Summary: Atoms and subatomic particles sometimes obey rules that aren't relevant to larger objects.

Key ideas

- ✪ Electrons occupy energy levels in atoms. In order to change energy levels, electrons must absorb or emit photons.

- ✪ The energy of a photon is given by $E = \dfrac{hc}{\lambda}$, where $hc = 1240$ eV·nm.

- ✪ Draw an energy level diagram with every atomic physics problem.
- ✪ Three types of nuclear decay can occur: Alpha, Beta, and Gamma.
- ✪ Charge and momentum are conserved in nuclear decay.
- ✪ Mass-energy is conserved in nuclear decay. The conversion between energy and mass is $E = mc^2$.

Relevant Equations

Energy of a photon:

$$E = hf = \frac{hc}{\lambda}$$

Momentum of a photon:

$$p = \frac{h}{\lambda} = \frac{E}{c}$$

de Broglie wavelength:

$$\lambda = \frac{h}{p}$$

Energy-mass conversion:

$$E = mc^2$$

Atomic and nuclear physics takes us into the bizarre world of the very, very small. This is the realm where the distinction between waves and particles gets fuzzy, where things without mass can have momentum, and where elements spontaneously break apart.

But even if the phenomena seem bizarre, the physics that describes them, fortunately, isn't. As far as the AP exam is concerned (and this, again, is a topic that appears exclusively on the Physics B exam), you won't be writing wacky equations or using lots of cryptic symbols. The math tends to be pretty straightforward, and, as we show you in this chapter, there really are only a few concepts you need to know to answer any question the AP exam could throw at you. So there's nothing to worry about . . . after all, it's only nuclear physics, right?

Subatomic Particles

We already introduced the key subatomic particles in Chapter 19, when we first discussed the concept of electric charge. But we review them briefly here.

Atoms, as you know, are the fundamental units that make up matter. An atom is composed of a small, dense core, called the **nucleus,** along with a larger, diffuse "cloud" of electrons that surrounds the nucleus. The nucleus of an atom is made up of two types of subatomic particles: protons and neutrons. Protons are positively charged particles, neutrons carry no charge, and electrons are negatively charged. For an atom to be electrically neutral, therefore, it must contain an equal number of protons and electrons. When an atom is not electrically neutral, it is called an **ion;** a positive ion has more protons than electrons, and a negative ion has more electrons than protons.

One particle we didn't introduce in Chapter 19 is the **photon.** A photon is a particle of light.[1] Photons have no mass, but they can transfer energy to or from electrons.

Table 24.1 summarizes the properties of the four subatomic particles. Note that the mass of a proton (and, similarly, the mass of a neutron) equals one **atomic mass unit,** abbreviated "1 amu." This is a convenient unit of mass to use when discussing atomic physics.

Table 24.1   Properties of the Subatomic Particles

NAME	MASS	CHARGE
Proton	1.67×10^{-27} kg = 1 amu	Positive
Neutron	1.67×10^{-27} kg = 1 amu	Zero
Electron	9.11×10^{-31} kg	Negative
Photon	0	Zero

---

[1] "Wait a minute—didn't the double-slit experiment prove that light is a wave?!?" This is one of the many times where the bizarreness of the atomic world emerges. Yes, light does behave like a wave. But light behaves like a particle also, as we see later in this chapter. It's not that light is a wave some of the time and a particle the rest of the time; rather, light is simply light all the time, and some of its properties are wave-like and some are particle-like.

The Electron-Volt

The **electron-volt**, abbreviated eV, is a unit of energy that's particularly useful for problems involving subatomic particles. One eV is equal to the amount of energy needed to change the potential of an electron by one volt. For example, imagine an electron nearby a positively charged particle, such that the electron's potential is 4 V. If you were to push the electron away from the positively charged particle until its potential was 5 V, you would need to use 1 eV of energy.

$$1 \text{ eV} = 1.6 \times 10^{-19} \text{ J}$$

The conversion to joules shows that an eV is an itty-bitty unit of energy. However, such things as electrons and protons *are* itty-bitty particles. So the electron-volt is actually a perfectly sized unit to use when talking about the energy of a subatomic particle.

Energy Levels in an Atom

The electrons in an atom can only have certain, specific amounts of energy. This is an unusual concept—if an electron is whirling around an atom with a given amount of energy, couldn't you just nudge it slightly to give it a tiny bit more energy? No, you can't. To understand what we mean, look at Figure 24.1.

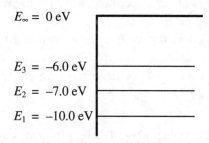

Figure 24.1   Energy levels in a hypothetical atom.

In Figure 24.1, we've drawn a few of the energy levels of a hypothetical atom. Let's start by looking at E_1. This is the lowest energy level, and it is called the **ground state energy** of the atom. When an electron is sitting as close to the nucleus as possible, and when it's completely unexcited, it will have the energy of E_1, -10 eV.

To increase the energy of the electron—to move the electron into a higher energy level—energy must somehow be transferred to the electron. This energy transfer is done by a photon. To jump up in energy, an electron absorbs a photon, and to drop down in energy, an electron emits a photon.

The energy diagram in Figure 24.1 tells us that to get to E_2 from the ground state, an electron must absorb a photon carrying 3 eV—and *only* 3 eV!—of energy. If a photon with an energy of 2.9 eV comes along and knocks into the electron, nothing will happen. Similarly, if a photon with an energy of 3.1 eV comes along, nothing will happen. It's all or nothing; either the electron gets just the right amount of energy to go from one energy level to the next, or it doesn't.

How does the electron get from the ground state to E_3? There are two ways. The electron could first absorb a photon with an energy of 3 eV, taking it from E_1 to E_2, and then it could absorb another photon with an energy of 1 eV, taking it from E_2 to E_3. Or, the electron could start in E_1 and simply absorb a photon with an energy of 4 eV.

We've been talking about photons having certain energies, but we haven't yet told you how to figure out the energy of a photon. Here's the formula:

$$\boxed{E = hf}$$

This formula tells us that the energy of a photon is equal to Planck's constant, h, which is 6.63×10^{-34} J·s (this value is given to you on the constants sheet), multiplied by the frequency of the photon. You should remember from Chapter 23 that the frequency of a wave is related to the wavelength by the formula

$$v = \lambda f$$

For light, the velocity is c, or 3×10^8 m/s, so we can instead write

$$c = \lambda f$$

This means that we can rewrite the equation for the energy of a photon to read

$$\boxed{E = \frac{hc}{\lambda}}$$

These formulas tell us that a photon with a high frequency, and therefore with a small wavelength, is higher in energy than a photon with a low frequency and long wavelength. So gamma rays, for example, are a lot higher energy than radio waves because gamma rays have a higher frequency.

Okay, let's try these formulas with a problem.

> What wavelength of light must an electron absorb in order to get from E_1 to E_2 in Figure 24.1?

This is a simple plug-and-chug problem. Figure 24.1 gives us the energy levels, and we just need to use our formula to find the wavelength of the absorbed photon. But what value do we use for the energy? −10 eV? −7 eV? Wrong . . . the value we use is the *energy of the jump*, not the energy of one of the states. To get from E_1 to E_2, an electron must absorb 3 eV of energy, so that's the value we use in the formula:

$$3 \text{ eV} = \frac{hc}{\lambda}$$

At this point, the problem is simply a matter of converting units and using your calculator. However, solving it is a lot easier if you know that the quantity hc equals 1240 eV·nm. This value is on the constants sheet, and it is *very* important to know! Look at how, by knowing this value, we can solve for the photon's wavelength very quickly.

$$\lambda = \frac{1240 \text{ eV} \cdot \text{nm}}{3 \text{ eV}}$$

$$\lambda = 410 \text{ nm}$$

In other words, for an electron in Figure 24.1 to go from the ground state to E_2, it must absorb light with a wavelength of exactly 410 nm, which happens to be a lovely shade of violet.

If you look back at Figure 24.1, you might wonder what's going on in the gap between E_3 and E_∞. In fact, this gap is likely filled with lots more energy levels—E_4, E_5, E_6 . . . However, as you go up in energy, the energy levels get squeezed closer and closer together (notice, for example, how the energy gap between E_1 and E_2 is greater than the gap between E_2 and E_3). However, we didn't draw all these energy levels, because our diagram would have become too crowded.

The presence of all these other energy levels, though, raises an interesting question. Clearly, an electron can keep moving from one energy level to the next if it absorbs the appropriate photons, getting closer and closer to E_∞. But can it ever get beyond E_∞? That is, if an electron in the ground state were to absorb a photon with an energy greater than 10 eV, what would happen?

The answer is that the electron would be ejected from the atom. It takes exactly 10 eV for our electron to escape the electric pull of the atom's nucleus—this minimum amount of energy needed for an electron to escape an atom is called the **ionization energy**[2]—so if the electron absorbed a photon with an energy of, say, 11 eV, then it would use 10 of those eV to escape the atom, and it would have 1 eV of leftover energy. That leftover energy takes the form of kinetic energy.

> An electron in the ground state of the atom depicted in Figure 24.1 absorbs a photon with an energy of 11 eV. The electron is ejected from the atom. What is the speed of the ejected electron?

As we said above, it takes 10 eV for our electron to escape the atom, which leaves 1 eV to be converted to kinetic energy. The formula for kinetic energy requires that we use values in standard units, so we need to convert the energy to joules.

$$1 \text{ eV}\left(1.6\times10^{-19}\ \frac{\text{J}}{\text{eV}}\right) = \tfrac{1}{2}mv^2$$

If we plug in the mass of an electron for m, we find that $v = 5.9 \times 10^5$ m/s. Is that a reasonable answer? It's certainly quite fast, but electrons generally travel very quickly. And it's several orders of magnitude slower than the speed of light, which is the fastest anything can travel. So our answer seems to make sense.

The observation that electrons, when given enough energy, can be ejected from atoms was the basis of one of the most important discoveries of twentieth century physics: the **photoelectric effect**.

> **Photoelectric Effect:** Energy in the form of light can cause an atom to eject one of its electrons, but only if the frequency of the light is above a certain value.

This discovery was surprising, because physicists in the early twentieth century thought that the brightness of light, and not its frequency, was related to the light's energy. But no matter how bright the light was that they used in experiments, they couldn't make atoms eject their

---

[2] The term **work function** refers to the energy necessary for an electron to escape from a metal surface—this is a very similar concept and is used in discussion of the photoelectric effect below.

electrons unless the light was above a certain frequency. This frequency became known, for obvious reasons, as the **cutoff frequency**. The cutoff frequency is different for every type of atom.

A metal surface has a work function of 10 eV. What is the cutoff frequency for this metal?

Remembering that hc equals 1240 eV·nm, we can easily find the wavelength of a photon with an energy of 10 eV:

$$\lambda = \frac{1240 \text{ eV} \cdot \text{nm}}{10 \text{ eV}}$$

$$\lambda = 124 \text{ nm}$$

Using the equation $c = \lambda f$, we find that $f = \textbf{2.42} \times \textbf{10}^{15} \textbf{ Hz}$. So any photon with a frequency equal to or greater than 2.42×10^{15} Hz carries enough energy to eject a photon from the metal surface.

Momentum of a Photon

Given that photons can "knock" an electron away from a nucleus, it might seem reasonable to talk about the momentum of a photon. Photons don't have mass, of course, but the momentum of a photon can nonetheless be found by this equation:

$$p = \frac{h}{\lambda}$$

The momentum of a photon equals Planck's constant divided by the photon's wavelength. You can rearrange this equation a bit to show that the energy of a photon, which equals hc/λ, also equals the photon's momentum times c.

de Broglie Wavelength

We've been discussing over the past few pages how light, which acts like a wave, also has the characteristics of particles. As it turns out, things with mass, which we generally treat like particles, also have the characteristics of waves. According to de Broglie, a moving mass behaves like a wave. The wavelength of a moving massive particle, called its **de Broglie wavelength**, is found by this formula:

$$\lambda = \frac{h}{p}$$

So an object's de Broglie wavelength equals Planck's constant divided by the object's momentum. For a massive particle, momentum $p = mv$. Note that the formula above is valid even for massless photons—it's just a rearrangement of the definition of the photon's momentum. What's interesting here is that the formula can assign a "wavelength" even to massive particles.

Okay, sure, massive particles behave like waves . . . so how come when you go for a jog, you don't start diffracting all over the place?

Let's figure out your de Broglie wavelength. We'll assume that your mass is approximately 60 kg and that your velocity when jogging is about 5 m/s. Plugging these values into de Broglie's equation, we find that your wavelength when jogging is about 10^{-36} m, which is *way* too small to detect.[3]

Only subatomic particles have de Broglie wavelengths large enough to actually be detected. Moving neutrons can be diffracted, but *you* can't.

Three Types of Nuclear Decay Processes

When physicists first investigated nuclear decay in the early twentieth century, they didn't quite know what they were seeing. The subatomic particles (protons, neutrons, electrons, and so forth) had not been definitively discovered and named. But, physicists did notice that certain kinds of particles emerged repeatedly from their experiments. They called these particles **alpha**, **beta**, and **gamma** particles.

Years later, physicists found out what these particles actually are:

- Alpha particle, α: two protons and two neutrons stuck together
- Beta particle, β: an electron or a positron[4]
- Gamma particle, γ: a photon

Be sure you recall the properties of these particles; if you have to, look in the glossary for a reminder. A couple of observations: The alpha particle is by far the most massive; the gamma particle is both massless and chargeless.

Nuclear Notation

The two properties of a nucleus that are most important are its **atomic number** and its **mass number**. The atomic number, Z, tells how many protons are in the nucleus; this number determines what element we're dealing with because each element has a unique atomic number. The mass number, A, tells the *total* number of nuclear particles a nucleus contains. It is equal to the atomic number plus the number of neutrons in the nucleus. **Isotopes** of an element have the same atomic number, but different mass numbers.

A nucleus is usually represented using the following notation:

$$^A_Z Symbol$$

where "symbol" is the symbol for the element we're dealing with. For example, $^4_2 He$ represents helium with two protons and two neutrons. A different isotope of helium might be $^5_2 He$, which contains three neutrons.

Now, you don't have to memorize the periodic table, nor do you have to remember what element is number 74.[5] But you *do* have to recognize what's wrong with this nucleus: $^{170}_{76} W$. Since we just told you in the footnote that tungsten is element 74, then an element with atomic number 76 *can't* be tungsten!

---

[3] The size of a proton is about 10^{-15} m, so your de Broglie wavelength is a hundred billion times smaller than a proton.

[4] The **positron** is the antimatter equivalent of an electron. It has the same mass as an electron, and the same amount of charge, but the charge is positive. Since there are two different types of beta particles, we often write them as β^+ for the positron and β^- for the electron.

[5] Tungsten: symbol "W." Used in light bulb filaments.

Alpha Decay

Alpha decay happens when a nucleus emits an alpha particle. Since an alpha particle has two neutrons and two protons, then the **daughter nucleus** (the nucleus left over after the decay) must have two fewer protons and two fewer neutrons than it had initially.

> $^{238}_{92}U$ undergoes alpha decay. Give the atomic number and mass number of the nucleus produced by the alpha decay.

The answer is found by simple arithmetic. The atomic number decreases by two, to **90**. The mass number decreases by four, to **234**. (The element formed is thorium, but you don't have to know that.) This alpha decay can be represented by the equation below:

$$^{238}_{92}U \rightarrow \, ^{234}_{90}Th + \, ^{4}_{2}\alpha$$

Beta Decay

In beta decay, a nucleus emits either a positron (β^+ decay) or an electron (β^- decay). Because an electron is not a normal nuclear particle, the total mass number of the nucleus must stay the same after beta decay. But the total charge present must not change—charge is a conserved quantity. So, consider an example of neon (Ne) undergoing β^+ decay:

$$^{19}_{10}Ne \rightarrow \, ^{19}_{9}F + e^+$$

Here e^+ indicates the positron. The mass number stayed the same. But look at the total charge present. Before the decay, the neon nucleus has a charge of +10. After the decay, the total charge must still be +10, and it is: +9 for the protons in the Fluorine (F), and +1 for the positron. Effectively, then, in β^+ decay, a proton turns into a neutron and emits a positron.

For β^- decay, a neutron turns into a proton, as in the decay process important for carbon dating:

$$^{14}_{6}C \rightarrow \, ^{14}_{7}N + e^-$$

The mass number of the daughter nucleus didn't change. But the total charge of the carbon nucleus was initially +6. Thus, a total charge of +6 has to exist after the decay, as well. This is accounted for by the electron (charge −1) and the nitrogen (charge +7).

In beta decay, a **neutrino** or an **antineutrino** also must be emitted to carry off some extra energy. But that doesn't affect the products of the decay, just the kinetic energy of the products.

Gamma Decay

A gamma particle is a photon, a particle of light. It has no mass and no charge. So a nucleus undergoing gamma decay does not change its outward appearance:

$$^{238}_{92}U \rightarrow \, ^{238}_{92}U + \gamma$$

However, the photon carries away some energy and momentum, so the nucleus recoils.

$E = mc^2$

KEY IDEA

We can imagine that the nuclei in the last few examples are at rest before emitting alpha, beta, or gamma particles. But when the nuclei decay, the alpha, beta, or gamma particles come whizzing out at an appreciable fraction of light speed.[6] So, the daughter nucleus must recoil after decay in order to conserve momentum.

But now consider energy conservation. Before decay, no kinetic energy existed. After the decay, both the daughter nucleus and the emitted particle have gobs of kinetic energy.[7] Where did this energy come from? Amazingly, it comes from mass.

The total mass present before the decay is very slightly greater than the total mass present in both the nucleus and the decay particle after the decay. That *slight* difference in mass is called the **mass defect**, often labeled Δm. This mass is destroyed and converted into kinetic energy.

How much kinetic energy? Use Einstein's famous equation to find out. Multiply the mass defect by the speed of light squared:

$$E = (\Delta m)c^2$$

And that is how much energy is produced.

› Practice Problems

Multiple Choice:

1. Which of the following lists types of electromagnetic radiation in order from least to greatest energy per photon?

   (A) red, green, violet, ultraviolet, infrared
   (B) ultraviolet, infrared, red, green, violet
   (C) red, green, violet, infrared, ultraviolet
   (D) infrared, red, green, violet, ultraviolet
   (E) ultraviolet, violet, green, red, infrared

Questions 2 and 3
In a nuclear reactor, uranium fissions into krypton and barium via the reaction

$$n + {}^{235}_{92}U \rightarrow {}^{141}_{56}Ba + Kr + 3n.$$

2. What are the mass number A and atomic number Z of the resulting krypton nucleus?

	A	Z
(A)	92	36
(B)	90	36
(C)	94	36
(D)	92	33
(E)	90	33

3. How much mass is converted into the kinetic energy of the resulting nuclei?

   (A) 1 amu
   (B) 2 amu
   (C) 3 amu
   (D) zero
   (E) much less than 1 amu

4. ${}^{15}_{8}O$ decays via β^+ emission. Which of the following is the resulting nucleus?

   (A) ${}^{15}_{9}F$
   (B) ${}^{16}_{8}O$
   (C) ${}^{15}_{7}N$
   (D) ${}^{16}_{7}N$
   (E) ${}^{16}_{9}F$

---

[6] 100% of the speed of light in the case of a gamma particle!

[7] No, the energy doesn't cancel out because the nucleus and the particle move in opposite directions; energy is a scalar, so direction is irrelevant.

Free Response:

$E = 0$

$E_2 = -1.2$ eV

$E_1 = -3.3$ eV

5. A hypothetical atom has two energy levels, as shown above.

(a) What wavelengths of electromagnetic radiation can be absorbed by this atom? Indicate which of these wavelengths, if any, represents visible light.

(b) Now, monochromatic 180-nm ultraviolet radiation is incident on the atom, ejecting an electron from the ground state. What will be
  i. the ejected electron's kinetic energy
  ii. the ejected electron's speed
  iii. the incident photon's speed

For parts (c) and (d), imagine that the 180-nm radiation ejected an electron that, instead of being in the ground state, was initially in the −1.2 eV state.

(c) Would the speed of the ejected electron increase, decrease, or stay the same? Justify your answer briefly.

(d) Would the speed of the incident photon increase, decrease, or stay the same? Justify your answer briefly.

› Solutions to Practice Problems

1. **D**—The radiation with the highest frequency (or shortest wavelength) has the highest energy per photon by $E = hf$. In the visible spectrum, red has the longest wavelength and violet has the shortest. Outside the visible spectrum, infrared radiation has a longer wavelength than anything visible, and ultraviolet has a shorter wavelength than anything visible. So, infrared has the smallest energy per photon, and so on up the spectrum to ultraviolet with the most energy per photon.

2. **A**—The total number of protons + neutrons is conserved. Before the reaction, we have one free neutron plus 235 protons and neutrons in the uranium, for a total of 236 amu. After the reaction, we have 141 amu in the barium plus 3 free neutrons for a total of 144 amu . . . leaving 92 AMUs for the krypton.

   Charge is also conserved. Before the reaction, we have a total charge of +92 from the protons in the uranium. After the reaction, we have 56 protons in the barium. Since a neutron carries no charge, the krypton must account for the remaining 36 protons.

3. **E**—Einstein's famous equation is written $\Delta E = mc^2$, because it is only the lost mass that is converted into energy. Since we still have a total of 236 amu after the reaction, an entire amu of mass was *not* converted to energy. Still, the daughter particles have kinetic energy because they move. That energy came from a very small mass difference, probably about a million times less than one amu.

4. **C**—In β^+ emission, a positron is ejected from the nucleus. This does not change the total number of protons + neutrons, so the atomic mass A is still 15. However, charge must be conserved. The original O nucleus had a charge of +8, so the products of the decay must combine to a charge of +8. These products are a positron, charge +1, and the daughter nucleus, which must have charge +7.

5. (a) $\Delta E = hc/\lambda$, so $hc/\Delta E = \lambda$. $hc = 1240$ eV·nm, as found on the equation sheet. ΔE represents the difference in energy levels. An electron in the ground state can make either of two jumps: it could absorb a 2.1-eV photon to get to the middle energy level, or it could absorb a 3.3 eV photon to escape the atom. An electron in the middle state could absorb a 1.2-eV photon to escape the atom. That makes three different energies. Convert these energies to wavelengths using $\Delta E = hc/\lambda$, so $hc/\Delta E = \lambda$. $hc = 1240$ eV·nm, as found on the equation sheet; ΔE represents the energy of the absorbed photon, listed above. These photons thus have wavelengths of 590 nm for the E_1 to E_2 transition; 380 nm or less for the E_1 to E_∞ transition; and 1030 nm or less for the E_2 to E_∞ transition. Only the 590 nm wavelength is visible because the visible spectrum is from about 400–700 nm.

(b) i. Find the energy of the incident photon using $\Delta E = hc/\lambda$, getting 6.9 eV. This is the total energy absorbed by the electron, but 3.3 eV of this is used to escape the atom. The remaining 3.6 eV is kinetic energy.

ii. To find speed, set kinetic energy equal to $\frac{1}{2}mv^2$. However, to use this formula, the kinetic energy must be in standard units of joules. Convert 3.6 eV to joules by multiplying by 1.6×10^{-19} J/eV (this conversion is on the constant sheet), getting a kinetic energy of 5.8×10^{-19} J. Now solve the kinetic energy equation for velocity, getting a speed of 1.1×10^6 m/s. This is reasonable—fast, but not faster than the speed of light.

iii. A photon is a particle of light. Unless it is in an optically dense material (which the photons here are not), the speed of a photon is always 3.0×10^8 m/s.

(c) The electron absorbs the same amount of energy from the incident photons. However, now it only takes 1.2 eV to get out of the atom, leaving 5.7 eV for kinetic energy. With a larger kinetic energy, the ejected electron's speed is greater than before.

(d) The speed of the photon is still the speed of light, 3.0×10^8 m/s.

› Rapid Review

- Atoms contain a nucleus, made of protons and neutrons, and one or more electrons that orbit that nucleus. Protons, neutrons, and electrons all have mass. By contrast, photons are subatomic particles without mass.

- The electron-volt is a unit of energy that's convenient to use when solving atomic physics problems.

- The electrons that surround an atom can only have certain, specific amounts of energy. To go from a low energy level to a high energy level, an electron absorbs a photon. To go from a high energy level to a low energy level, an electron emits a photon.

- If an electron absorbs a photon that has a higher energy than the electron's work function, the electron will be expelled from the atom.

- Moving particles have a characteristic wavelength, found by the de Broglie equation.

- Nuclei can undergo three types of decay. In alpha decay, a nucleus emits an alpha particle, which consists of two protons and two neutrons. In beta decay, a nucleus emits either a positron or an electron. In gamma decay, a nucleus emits a photon. When solving nuclear decay problems, remember to conserve both mass and charge.

- During nuclear decay, mass is converted to energy. The relationship between the mass defect and the gained energy is found by Einstein's famous formula, $E = (\Delta m)c^2$.

STEP **5**

Build Your Test-Taking Confidence

Physics B—Practice Exam—Multiple-Choice Questions

ANSWER SHEET

1 (A) (B) (C) (D) (E)   26 (A) (B) (C) (D) (E)   51 (A) (B) (C) (D) (E)
2 (A) (B) (C) (D) (E)   27 (A) (B) (C) (D) (E)   52 (A) (B) (C) (D) (E)
3 (A) (B) (C) (D) (E)   28 (A) (B) (C) (D) (E)   53 (A) (B) (C) (D) (E)
4 (A) (B) (C) (D) (E)   29 (A) (B) (C) (D) (E)   54 (A) (B) (C) (D) (E)
5 (A) (B) (C) (D) (E)   30 (A) (B) (C) (D) (E)   55 (A) (B) (C) (D) (E)
6 (A) (B) (C) (D) (E)   31 (A) (B) (C) (D) (E)   56 (A) (B) (C) (D) (E)
7 (A) (B) (C) (D) (E)   32 (A) (B) (C) (D) (E)   57 (A) (B) (C) (D) (E)
8 (A) (B) (C) (D) (E)   33 (A) (B) (C) (D) (E)   58 (A) (B) (C) (D) (E)
9 (A) (B) (C) (D) (E)   34 (A) (B) (C) (D) (E)   59 (A) (B) (C) (D) (E)
10 (A) (B) (C) (D) (E)   35 (A) (B) (C) (D) (E)   60 (A) (B) (C) (D) (E)
11 (A) (B) (C) (D) (E)   36 (A) (B) (C) (D) (E)   61 (A) (B) (C) (D) (E)
12 (A) (B) (C) (D) (E)   37 (A) (B) (C) (D) (E)   62 (A) (B) (C) (D) (E)
13 (A) (B) (C) (D) (E)   38 (A) (B) (C) (D) (E)   63 (A) (B) (C) (D) (E)
14 (A) (B) (C) (D) (E)   39 (A) (B) (C) (D) (E)   64 (A) (B) (C) (D) (E)
15 (A) (B) (C) (D) (E)   40 (A) (B) (C) (D) (E)   65 (A) (B) (C) (D) (E)
16 (A) (B) (C) (D) (E)   41 (A) (B) (C) (D) (E)   66 (A) (B) (C) (D) (E)
17 (A) (B) (C) (D) (E)   42 (A) (B) (C) (D) (E)   67 (A) (B) (C) (D) (E)
18 (A) (B) (C) (D) (E)   43 (A) (B) (C) (D) (E)   68 (A) (B) (C) (D) (E)
19 (A) (B) (C) (D) (E)   44 (A) (B) (C) (D) (E)   69 (A) (B) (C) (D) (E)
20 (A) (B) (C) (D) (E)   45 (A) (B) (C) (D) (E)   70 (A) (B) (C) (D) (E)
21 (A) (B) (C) (D) (E)   46 (A) (B) (C) (D) (E)
22 (A) (B) (C) (D) (E)   47 (A) (B) (C) (D) (E)
23 (A) (B) (C) (D) (E)   48 (A) (B) (C) (D) (E)
24 (A) (B) (C) (D) (E)   49 (A) (B) (C) (D) (E)
25 (A) (B) (C) (D) (E)   50 (A) (B) (C) (D) (E)

Physics B—Practice Exam—Multiple-Choice Questions

Time: 90 minutes. You may refer to the Constants sheet found in the Appendixes. However, you may not use the Equations sheet, and you may not use a calculator on this portion of the exam.

1. A ball is thrown off of a 25-m-high cliff. Its initial velocity is 25 m/s, directed at an angle of 53° above the horizontal. How much time elapses before the ball hits the ground? (sin 53° = 0.80; cos 53° = 0.60; tan 53° = 1.3)

   (A) 3.0 s
   (B) 5.0 s
   (C) 7.0 s
   (D) 9.0 s
   (E) 11.0 s

2. A ball is dropped off of a cliff of height h. Its velocity upon hitting the ground is v. At what height above the ground is the ball's velocity equal to $v/2$?

   (A) $\dfrac{h}{4}$

   (B) $\dfrac{h}{2}$

   (C) $\dfrac{h}{\sqrt{2}}$

   (D) $\dfrac{h}{8}$

   (E) $\dfrac{h}{2\sqrt{2}}$

3. A golf cart moves with moderate speed as it reaches the base of a short but steep hill. The cart coasts up the hill (without using its brake or gas pedal). At the top of the hill the cart just about comes to rest; but then the cart starts to coast down the other side of the hill. Consider the forward motion of the cart to be positive. Which of the following velocity–time graphs best represents the motion of the cart?

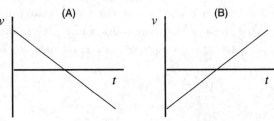

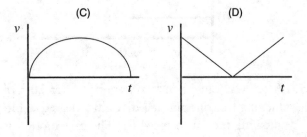

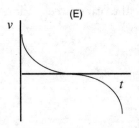

GO ON TO THE NEXT PAGE

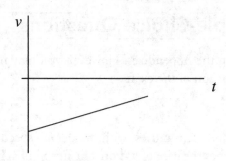

4. The velocity–time graph above represents a car on a freeway. North is defined as the positive direction. Which of the following describes the motion of the car?

(A) The car is traveling north and slowing down.
(B) The car is traveling south and slowing down.
(C) The car is traveling north and speeding up.
(D) The car is traveling south and speeding up.
(E) The car is traveling northeast and speeding up.

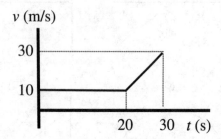

5. A car was caught in heavy traffic. After 20 s of moving at constant speed, traffic cleared a bit, allowing the car to speed up. The car's motion is represented by the velocity–time graph above. What was the car's acceleration while it was speeding up?

(A) 0.5 m/s^2
(B) 1.0 m/s^2
(C) 1.5 m/s^2
(D) 2.0 m/s^2
(E) 3.0 m/s^2

6. A block of mass m sits on the ground. A student pulls up on the block with a tension T, but the block remains in contact with the ground. What is the normal force on the block?

(A) $T + mg$
(B) $T - mg$
(C) mg
(D) $mg - T$
(E) T

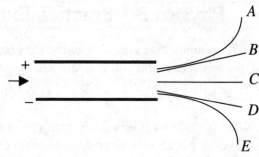

7. A proton moving at constant velocity enters the region between two charged plates, as shown above. Which of the paths shown correctly indicates the proton's trajectory after leaving the region between the charged plates?

(A) A
(B) B
(C) C
(D) D
(E) E

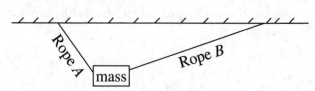

8. A mass hangs from two ropes at unequal angles, as shown above. Which of the following makes correct comparisons of the horizontal and vertical components of the tension in each rope?

	Horizontal tension	Vertical tension
(A)	greater in rope B	greater in rope B
(B)	equal in both ropes	greater in rope A
(C)	greater in rope A	greater in rope A
(D)	equal in both ropes	equal in both ropes
(E)	greater in rope B	equal in both ropes

9. A free-body diagram includes vectors representing the individual forces acting on an object. Which of these quantities should NOT appear on a free-body diagram?

(A) tension of a rope
(B) mass times acceleration
(C) kinetic friction
(D) static friction
(E) weight

GO ON TO THE NEXT PAGE

10. An object rolls along level ground to the right at constant speed. Must there be any forces pushing this object to the right?

(A) Yes: the *only* forces that act must be to the right.
(B) Yes: but there could also be a friction force acting to the left.
(C) No: no forces can act to the right.
(D) No: while there can be forces acting, no force MUST act.
(E) The answer depends on the speed of the object.

11. A person stands on a scale in an elevator. He notices that the scale reading is lower than his usual weight. Which of the following could possibly describe the motion of the elevator?

(A) It is moving down at constant speed.
(B) It is moving down and slowing down.
(C) It is moving up and slowing down.
(D) It is moving up and speeding up.
(E) It is moving up with constant speed.

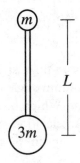

12. A mass *m* is attached to a mass 3*m* by a rigid bar of negligible mass and length *L*. Initially, the smaller mass is located directly above the larger mass, as shown above. How much work is necessary to flip the rod 180° so that the larger mass is directly above the smaller mass?

(A) $4mgL$
(B) $2mgL$
(C) mgL
(D) $4\pi mgL$
(E) $2\pi mgL$

13. A ball rolls horizontally with speed *v* off of a table a height *h* above the ground. Just before the ball hits the ground, what is its speed?

(A) $\sqrt{2gh}$

(B) $v\sqrt{2gh}$

(C) $\sqrt{v^2 + 2gh}$

(D) v

(E) $v + \sqrt{2gh}$

Questions 14 and 15

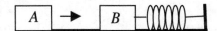

Block *B* is at rest on a smooth tabletop. It is attached to a long spring, which in turn is anchored to the wall. Block *A* slides toward and collides with block *B*. Consider two possible collisions:

Case I: Block *A* bounces back off of block *B*.
Case II: Block *A* sticks to block *B*.

14. Which of the following is correct about the speed of block *B* immediately after the collision?

(A) It is faster in case II than in case I ONLY if block *B* is heavier.
(B) It is faster in case I than in case II ONLY if block *B* is heavier.
(C) It is faster in case II than in case I regardless of the mass of each block.
(D) It is faster in case I than in case II regardless of the mass of each block.
(E) It is the same in either case regardless of the mass of each block.

15. Which is correct about the period of the ensuing oscillations after the collision?

(A) The period is greater in case II than in case I if block *B* is heavier.
(B) The period is greater in case I than in case II if block *B* is heavier.
(C) The period is greater in case II than in case I regardless of the mass of each block.
(D) The period is greater in case I than in case II regardless of the mass of each block.
(E) The period is the same in either case.

GO ON TO THE NEXT PAGE

16. A ball collides with a stationary block on a frictionless surface. The ball sticks to the block. Which of the following would NOT increase the force acting on the ball during the collision?

(A) increasing the time it takes the ball to change the block's speed
(B) arranging for the ball to bounce off of the block rather than stick
(C) increasing the speed of the ball before it collides with the block
(D) increasing the mass of the block
(E) anchoring the block in place so that the ball/block combination cannot move after collision

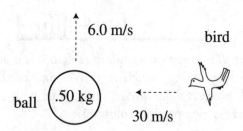

17. A 0.30-kg bird is flying from right to left at 30 m/s. The bird collides with and sticks to a 0.50-kg ball that is moving straight up with speed 6.0 m/s. What is the magnitude of the momentum of the ball/bird combination immediately after collision?

(A) 12.0 N·s
(B) 9.5 N·s
(C) 9.0 N·s
(D) 6.0 N·s
(E) 3.0 N·s

18. A car slows down on a highway. Its engine is providing a forward force of 1000 N; the force of friction is 3000 N. It takes 20 s for the car to come to rest. What is the car's change in momentum during these 20 s?

(A) 10,000 kg·m/s
(B) 20,000 kg·m/s
(C) 30,000 kg·m/s
(D) 40,000 kg·m/s
(E) 60,000 kg·m/s

19. Which of the following quantities is NOT a vector?

(A) magnetic field
(B) electric force
(C) electric current
(D) electric field
(E) electric potential

20. Which of the following must be true of an object in uniform circular motion?

(A) Its velocity must be constant.
(B) Its acceleration and its velocity must be in opposite directions.
(C) Its acceleration and its velocity must be perpendicular to each other.
(D) It must experience a force away from the center of the circle.
(E) Its acceleration must be negative.

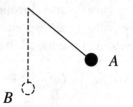

21. A ball of mass m on a string swings back and forth to a maximum angle of 30° to the vertical, as shown above. Which of the following vectors represents the acceleration, a, of the mass at point A, the highest point in the swing?

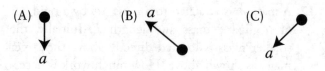

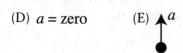

22. A planet of mass m orbits in a circle around a sun. The speed of the planet in its orbit is v; the distance from the planet to the sun is d. What is the magnitude and direction of the net force experienced by the planet?

    (A) $\dfrac{v^2}{d}$, toward the sun.

    (B) $m\dfrac{v^2}{d}$, toward the sun.

    (C) $m\dfrac{v^2}{d}$, away from the sun.

    (D) $m\dfrac{v^2}{d}$, along the orbital path.

    (E) $\dfrac{v^2}{d}$, along the orbital path.

23. A mass m on a spring oscillates on a horizontal surface with period T. The total mechanical energy contained in this oscillation is E. Imagine that instead a new mass $4m$ oscillates on the same spring with the same amplitude. What is the new period and total mechanical energy?

	Period	Total Mechanical Energy
(A)	T	E
(B)	$2T$	E
(C)	$2T$	$2E$
(D)	T	$4E$
(E)	$2T$	$16E$

24. A satellite orbits the moon in a circle of radius R. If the satellite must double its speed but maintain a circular orbit, what must the new radius of its orbit be?

    (A) $2\,R$
    (B) $4\,R$
    (C) $\frac{1}{2}\,R$
    (D) $\frac{1}{4}\,R$
    (E) R

25. The Space Shuttle orbits 300 km above the Earth's surface; the Earth's radius is 6400 km. What is the gravitational acceleration experienced by the Space Shuttle?

    (A) 4.9 m/s^2
    (B) 8.9 m/s^2
    (C) 9.8 m/s^2
    (D) 10.8 m/s^2
    (E) zero

26. A cube of ice (specific gravity 0.90) floats in a cup of water. Several hours later, the ice cube has completely melted into the glass. How does the water level after melting compare to the initial water level?

    (A) The water level is 10% higher after melting.
    (B) The water level is 90% higher after melting.
    (C) The water level is unchanged after melting.
    (D) The water level is 10% lower after melting.
    (E) The water level is 90% lower after melting.

27. A strong hurricane may include 50 m/s winds. Consider a building in such a hurricane. If the air inside the building is kept at standard atmospheric pressure, how will the outside air pressure compare to the inside air pressure?

    (A) Outside pressure will be the same.
    (B) Outside pressure will be about 2% greater.
    (C) Outside pressure will be about 2% lower.
    (D) Outside pressure will be about twice inside pressure.
    (E) Outside pressure will be about half inside pressure.

28. A heavy block sits on the bottom of an aquarium. Which of the following must be correct about the magnitude of the normal force exerted on the block by the aquarium bottom?

    (A) The normal force is equal to the block's weight.
    (B) The normal force is less than the block's weight.
    (C) The normal force is greater than the block's weight.
    (D) The normal force is equal to the buoyant force on the block.
    (E) The normal force is greater than the buoyant force on the block.

GO ON TO THE NEXT PAGE

29. The density of water near the ocean's surface is $\rho_o = 1.025 \times 10^3$ kg/m³. At extreme depths, though, ocean water becomes slightly more dense. Let ρ represent the density of ocean water at the bottom of a trench of depth d. What can be said about the gauge pressure at this depth?

(A) The pressure is greater than $\rho_o gd$, but less than ρgd.
(B) The pressure is greater than ρgd.
(C) The pressure is less than $\rho_o gd$.
(D) The pressure is greater than $\rho_o gd$, but could be less than or greater than ρgd.
(E) The pressure is less than ρgd, but could be less than or greater than $\rho_o gd$.

30. A 1.0-m-long brass pendulum has a period of 2.0 s on a very cold (−10°C) day. On a very warm day, when the temperature is 30°C, what is the period of this pendulum?

(A) 1.0 s
(B) 1.4 s
(C) 1.8 s
(D) 2.0 s
(E) 4.0 s

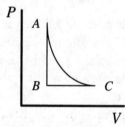

31. The state of a gas in a cylinder is represented by the *PV* diagram shown above. The gas can be taken through either the cycle *ABCA*, or the reverse cycle *ACBA*. Which of the following statements about the work done on or by the gas is correct?

(A) In both cases, the same amount of net work is done *by* the gas.
(B) In both cases, the same amount of net work is done *on* the gas.
(C) In cycle *ABCA* net work is done *on* the gas; in cycle *ACBA* the same amount of net work is done *by* the gas.
(D) In cycle *ABCA* net work is done *by* the gas; in cycle *ACBA* the same amount of net work is done *on* the gas.
(E) In both cycles net work is done *by* the gas, but more net work is done by the gas in *ACBA* than in *ABCA*.

32. At room temperature, the rms speed of nitrogen molecules is about 500 m/s. By what factor must the absolute temperature T of the Earth change for the rms speed of nitrogen to reach escape velocity, 11 km/s?

(A) T must decrease by a factor of 50.
(B) T must increase by a factor of 22.
(C) T must increase by a factor of 50.
(D) T must increase by a factor of 200.
(E) T must increase by a factor of 500.

33. One mole of He (atomic mass 4 amu) occupies a volume of 0.022 m³ at room temperature and atmospheric pressure. How much volume is occupied by one mole of O_2 (atomic mass 32 amu) under the same conditions?

(A) 0.022 m³
(B) 0.088 m³
(C) 0.176 m³
(D) 0.352 m³
(E) 0.003 m³

34. On average, how far apart are N_2 molecules at room temperature?

(A) 10^{-21} m
(B) 10^{-15} m
(C) 10^{-9} m
(D) 10^{-3} m
(E) 1 m

35. Which of the following is NOT a statement or consequence of the second law of thermodynamics, the law dealing with entropy?

(A) The net work done by a gas cannot be greater than the sum of its loss of internal energy and the heat added to it.
(B) Even an ideal heat engine cannot be 100% efficient.
(C) A warm ball of putty cannot spontaneously cool off and rise.
(D) A system cannot become more ordered unless some net work is done to obtain that order.
(E) Heat does not flow naturally from a low to high temperature.

GO ON TO THE NEXT PAGE

36. Experimenter A uses a very small test charge q_0, and experimenter B uses a test charge $2q_0$ to measure an electric field produced by stationary charges. A finds a field that is

    (A) greater than the field found by B
    (B) the same as the field found by B
    (C) less than the field found by B
    (D) either greater or less than the field found by B, depending on the accelerations of the test charges
    (E) either greater or less than the field found by B, depending on the masses of the test charges

37. Two isolated particles, A and B, are 4 m apart. Particle A has a net charge of $2Q$, and B has a net charge of Q. The ratio of the magnitude of the electrostatic force on A to that on B is

    (A) 4:1
    (B) 2:1
    (C) 1:1
    (D) 1:2
    (E) 1:4

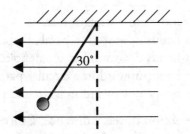

38. A uniform electric field points right to left. A small metal ball charged to +2 mC hangs at a 30° angle from a string of negligible mass, as shown above. The tension in the string is measured to be 0.1 N. What is the magnitude of the electric field? ($\sin 30° = 0.50$; $\cos 30° = 0.87$; $\tan 30° = 0.58$).

    (A)   25 N/C
    (B)   50 N/C
    (C)   2500 N/C
    (D)   5000 N/C
    (E) 10,000 N/C

39. 1.0 nC is deposited on a solid metal sphere of diameter 0.30 m. What is the magnitude of the electric field at the center of the sphere?

    (A) zero
    (B) 25 N/C
    (C) 100 N/C
    (D) 200 N/C
    (E) 400 N/C

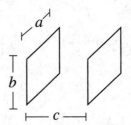

40. The parallel plate capacitor above consists of identical rectangular plates of dimensions $a \times b$, separated by a distance c. To cut the capacitance of this capacitor in half, which of these quantities should be doubled?

    (A) a
    (B) b
    (C) c
    (D) ab
    (E) abc

41. Two identical capacitors are connected in parallel to an external circuit. Which of the following quantities must be the same for both capacitors?

    I. the charge stored on the capacitor
    II. the voltage across the capacitor
    III. the capacitance of the capacitor
    (A) I only
    (B) II only
    (C) II and III only
    (D) I and III only
    (E) I, II, and III

GO ON TO THE NEXT PAGE

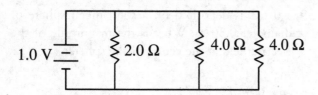

42. Three resistors are connected to a 1.0 V battery, as shown in the diagram above. What is the current through the 2.0 Ω resistor?

(A) 0.25 A
(B) 0.50 A
(C) 1.0 A
(D) 2.0 A
(E) 4.0 A

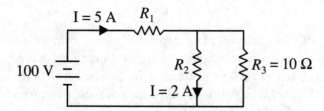

43. What is the voltage drop across R_3 in the circuit diagrammed above?

(A)   10 V
(B)   20 V
(C)   30 V
(D)   50 V
(E)   100 V

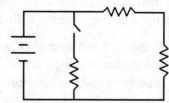

44. Three resistors are connected to an ideal battery as shown in the diagram above. The switch is initially open. When the switch is closed, what happens to the total voltage, current, and resistance in the circuit?

	Voltage	Current	Resistance
(A)	increases	increases	increases
(B)	does not change	does not change	does not change
(C)	does not change	decreases	increases
(D)	does not change	increases	decreases
(E)	decreases	decreases	decreases

45. On which of the following physics principles does Kirchoff's loop rule rest?

(A) conservation of charge
(B) conservation of mass
(C) conservation of energy
(D) conservation of momentum
(E) conservation of angular momentum

Questions 46 and 47

```
X X X X X  B
X X X X X
● → v X X X
X X X X X
X X X X
```

A uniform magnetic field B is directed into the page. An electron enters this field with initial velocity v to the right.

46. Which of the following best describes the path of the electron while it is still within the magnetic field?

(A) It moves in a straight line.
(B) It bends upward in a parabolic path.
(C) It bends downward in a parabolic path.
(D) It bends upward in a circular path.
(E) It bends downward in a circular path.

47. The electron travels a distance d, measured along its path, before exiting the magnetic field. How much work is done on the electron by the magnetic field?

(A) $evBd$
(B) zero
(C) $-evBd$
(D) $evBd \sin (d/2\pi)$
(E) $-evBd \sin (d/2\pi)$

GO ON TO THE NEXT PAGE

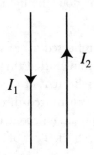

48. The circular wire shown above carries a current I in the counterclockwise direction. What will be the direction of the magnetic field at the center of the wire?

(A) into the page
(B) out of the page
(C) down
(D) up
(E) counterclockwise

49. Two parallel wires carry currents in opposite directions, as shown above. In what direction will the right-hand wire (the wire carrying current I_2) experience a force?

(A) left
(B) right
(C) into the page
(D) out of the page
(E) The right-hand wire will experience no force.

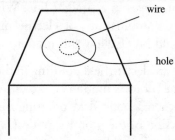

50. A loop of wire surrounds a hole in a table, as shown above. A bar magnet is dropped, north end down, from above the table through the hole. Let the positive direction of current be defined as counterclockwise as viewed from above. Which of the following graphs best represents the induced current I in the loop?

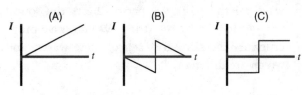

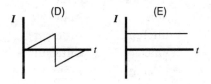

51. A rectangular loop of wire has dimensions $a \times b$ and includes a resistor R. This loop is pulled with speed v from a region of no magnetic field into a uniform magnetic field B pointing through the loop, as shown above. What is the magnitude and direction of the current through the resistor?

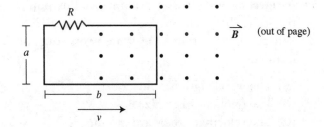

(A) Bav/R, left-to-right
(B) Bbv/R, left-to-right
(C) Bav/R, right-to-left
(D) Bbv/R, right-to-left
(E) Bba/R, right-to-left

52. A proton moves in a straight line. Which of the following combinations of electric or magnetic fields could NOT allow this motion?

    (A) only an electric field pointing in the direction of the proton's motion
    (B) only a magnetic field pointing opposite the direction of the proton's motion
    (C) an electric field and a magnetic field, each pointing perpendicular to the proton's motion
    (D) only a magnetic field pointing perpendicular to the proton's motion
    (E) only a magnetic field pointing in the direction of the proton's motion

53. A guitar string is plucked, producing a standing wave on the string. A person hears the sound wave generated by the string. Which wave properties are the same for each of these waves, and which are different?

	Wavelength	Velocity	Frequency
(A)	same	different	same
(B)	same	same	same
(C)	same	different	different
(D)	different	same	different
(E)	different	different	same

54. A traveling wave passes a point. At this point, the time between successive crests is 0.2 s. Which of the following statements can be justified?

    (A) The wavelength is 5 m.
    (B) The frequency is 5 Hz.
    (C) The velocity of the wave is 5 m/s.
    (D) The wavelength is 0.2 m.
    (E) The wavelength is 68 m.

55. What is the frequency of sound waves produced by a string bass whose height is 2 m? The speed of waves on the string is 200 m/s.

    (A) 5 Hz
    (B) 0.05 Hz
    (C) 50 Hz
    (D) 500 Hz
    (E) 5000 Hz

56. What property of a light wave determines the color of the light?

    (A) frequency
    (B) wavelength
    (C) velocity
    (D) amplitude
    (E) the medium through which the light wave travels

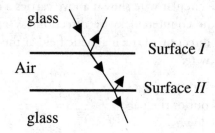

57. Light traveling through glass hits a thin film of air, as shown above. For which of the following beams of light does the light wave change phase by 180°?

    (A) the light reflected off of surface *I*
    (B) the light transmitted through surface *I*
    (C) the light reflected off of surface *II*
    (D) the light transmitted through surface *II*
    (E) both the light reflected off of surface *I* and the light reflected off of surface *II*

GO ON TO THE NEXT PAGE

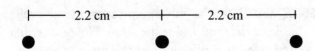

58. Light from a coherent source passes through a diffraction grating, producing an interference pattern on a screen. Three of the bright spots produced on the screen are 2.2 cm away from one another, as shown above. Now, a new diffraction grating is substituted, whose distance between lines is half of the original grating's. Which of the following shows the new interference pattern?

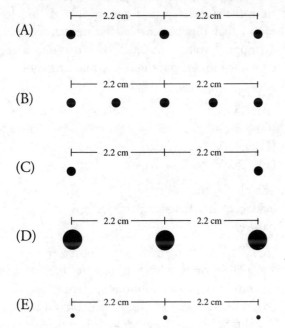

59. Which colors of visible light have the *largest* frequency, wavelength, and energy per photon?

	Frequency	Wavelength	Energy per photon
(A)	red	violet	violet
(B)	red	violet	red
(C)	violet	red	violet
(D)	violet	red	red
(E)	violet	violet	violet

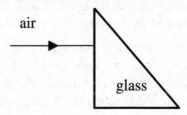

60. White light is incident on the triangular glass prism shown above. Why is it that blue and red light can be seen separately when the light exits the prism?

(A) The blue light speeds up more inside the glass.
(B) Some of the red light reflects at each interface.
(C) The light changes its frequency within the glass.
(D) The blue light bends farther away from normal at the left-hand interface.
(E) The blue light bends farther away from normal at the right-hand interface.

61. A convex lens projects a clear, focused image of a candle onto a screen. The screen is located 30 cm away from the lens; the candle sits 20 cm from the lens. Later, it is noticed that this same lens can also project a clear, focused image of the lighted windows of a building. If the building is located 60 meters away, what is the distance between the lens and the image of the building?

(A) 12 cm
(B) 24 cm
(C) 40 cm
(D) 80 cm
(E) 90 cm

62. Which of the optical instruments listed below can produce a virtual image of an object that is smaller than the object itself?

   I.   concave mirror
  II.   convex mirror
 III. concave lens
 IV. convex lens

    (A) I only
    (B) II only
    (C) III only
    (D) II and III only
    (E) I and IV only

GO ON TO THE NEXT PAGE

63. Which of the following does NOT describe a ray that can be drawn for a concave mirror?

(A) an incident ray through the mirror's center, reflecting right back through the center
(B) an incident ray through the center point, reflecting through the focal point
(C) an incident ray through the focal point, reflecting parallel to the principal axis
(D) an incident ray parallel to the principal axis, reflecting through the focal point
(E) an incident ray to the intersection of the principal axis with the mirror, reflecting at an equal angle.

64. Monochromatic light is incident on a photoelectric surface with work function W. The intensity, I, of the incident light is gradually increased. Which of the following graphs could represent the kinetic energy, KE, of electrons ejected by the photoelectric surface as a function of the intensity?

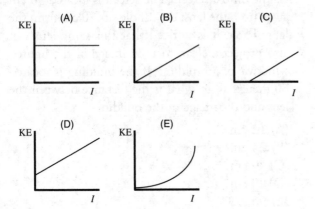

Questions 65 and 66

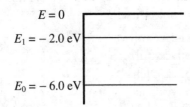

The energy levels of a hypothetical atom are shown above.

65. What wavelength of light must be absorbed to excite electrons from E_0 to E_1?

(A) 1240 nm
(B) 620 nm
(C) 100 nm
(D) 210 nm
(E) 310 nm

66. E_0 and E_1 are the only atomic energy levels available to electrons inside the atom. How many different colored photons can this atom emit when it captures an electron with no kinetic energy?

(A) zero
(B) one
(C) two
(D) three
(E) four

67. Monochromatic light is incident on a photoelectric surface of work function 3.5 eV. Electrons ejected from the surface create a current in a circuit. It is found that this current can be neutralized using a stopping voltage of 1.0 V. What is the energy contained in one photon of the incident light?

(A) 1.0 eV
(B) 2.5 eV
(C) 3.5 eV
(D) 4.5 eV
(E) 5.5 eV

68. Which of the following observations provides evidence that massive particles can have a wave nature?

(A) When burning hydrogen is observed through a spectrometer, several discrete lines are seen rather than a continuous spectrum.
(B) When a beam of electrons is passed through slits very close together and then projected on to a phosphorescent screen, several equally spaced bright spots are observed.
(C) When waves on the surface of the ocean pass a large rock, the path of the waves is bent.
(D) When alpha particles are passed through gold foil, most of the alpha particles go through the foil undeflected.
(E) When light is reflected off of a thin film, bright and dark fringes are observed on the film.

69. In which of the following nuclear processes do both the atomic number Z as well as the atomic mass A remain unchanged by the process?

(A) uranium fission
(B) alpha decay
(C) beta$^+$ decay
(D) beta$^-$ decay
(E) gamma decay

GO ON TO THE NEXT PAGE

70. A thorium nucleus emits an alpha particle. Which of the following fundamental physics principles can be used to explain why the direction of the daughter nucleus's recoil must be in the opposite direction of the alpha emission?

   I. Newton's third law

  II. conservation of momentum

III. conservation of energy

    (A) II only

    (B) III only

    (C) I and II only

    (D) II and III only

    (E) I, II, and III

STOP. End of Physics B—Practice Exam—Multiple-Choice Questions

Physics B—Practice Exam—Free-Response Questions

Time: 90 minutes. You may refer to the Constants sheet and Equations sheet in the Appendixes during this portion of the exam. You may also use a calculator during this portion of the exam.

(**Note:** In some years, the Physics B free-response section will be reduced to six rather than seven questions, still with a 90-minute time frame. We'll leave the seventh question in there for you for extra practice.)

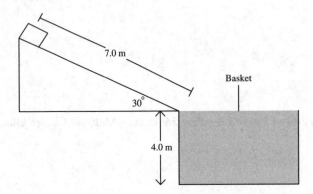

1. (15 points)

A package is at rest on top of a frictionless, 7.0 m long, 30° inclined plane, as shown above. The package slides down the plane, then drops 4.0 vertical meters into a waiting basket.

(a) How long does it take the package to reach the bottom of the plane?
(b) What is the magnitude and direction of the velocity of the package when it reaches the bottom of the plane?
(c) What is the TOTAL time it takes for the package both to slide down the plane and then to hit the bottom of the basket?
(d) How far, horizontally, from the left-hand edge of the basket will the package hit the bottom?
(e) If the basket were 8.0 m deep rather than 4.0 m deep, how would the answer to part (d) change? Check one box, and justify your answer without reference to an explicit numerical calculation.
  ☐ The horizontal distance would be doubled.
  ☐ The horizontal distance would be more than doubled.
  ☐ The horizontal distance would be increased, but less than doubled.

GO ON TO THE NEXT PAGE

2. (15 points)

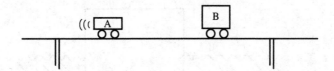

Two carts with different masses collide on a level track on which friction is negligible. Cart B is at rest before the collision. After the collision, the carts bounce off of one another; cart A moves backward, while cart B goes forward.

Consider these definitions of variables:

- m_A, the mass of cart A
- m_B, the mass of cart B
- v_A, the speed of cart A before collision
- v_A', the speed of cart A after collision
- v_B', the speed of cart B after collision

(a) Using only the variables listed above, derive an expression for the fraction of momentum conserved in this collision.

(b) Using only the variables listed above, derive an expression for the fraction of kinetic energy conserved in the collision.

(c) Which of the expressions you derived in parts (a) and (b) should always be equal to 100%? Justify your answer.

(d) You are given the following equipment:

    one sonic motion detector
    one stopwatch
    one meter stick

Describe a procedure that would allow an experimenter to measure the speeds of the carts before and after collision (keeping in mind that cart B is at rest before collision). Remember to be explicit about the placement of the motion detector; explain clearly how each speed can be determined from your measurements.

GO ON TO THE NEXT PAGE

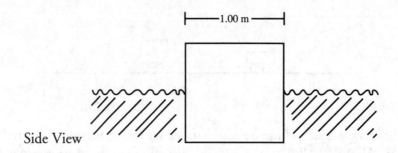

Side View

3. (15 points)

A large cube with a side of length 1.00 m is floating on top of a lake. Resting undisturbed, it floats exactly half submerged, as shown above. The density of water is 1000 kg/m³.

(a) Determine the mass of the cube.

A crane pushes the cube down into the water until it is ¾ underwater, as shown; the cube is then released from rest.

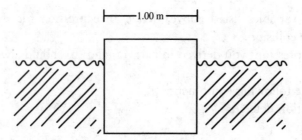

(b) Determine the magnitude and direction of the net force on the cube at the moment when it is released.
(c) After release, the cube oscillates up and down in simple harmonic motion, just as if it were attached to an ideal spring of "effective spring constant" *k*. Determine this effective spring constant.
(d) Determine the frequency of the resulting harmonic motion.
(e) If instead, the block is released after being pushed ⅘ of the way into the water, which of the following is correct about the frequency of the new harmonic motion? Check one of the three boxes below, then justify your answer.
    ☐ The frequency will be greater than before.
    ☐ The frequency will be smaller than before.
    ☐ The frequency will be the same as it was before.

GO ON TO THE NEXT PAGE

4. (15 points)

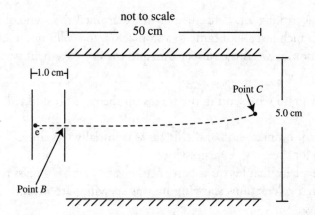

An electron starts from rest and is accelerated between two metal plates over a distance of 1.0 cm through a potential difference of 500 V. The electron exits the plates at point *B*, as shown in the diagram above. Neglect any relativistic effects.

(a) What is the velocity of this electron once it reaches point *B*?

(b) Once the electron has left point *B*, it enters directly between a second set of 50-cm-long parallel plates. These are arranged horizontally, separated by 5 cm, and with a 100 V potential difference between them.

    i. What are the magnitude and direction of the electric force, F_E, on the electron?

    ii. Determine the time it takes for the electron to travel the 50 cm through these plates to point *C*.

    iii. Determine the speed of the electron at point *C*.

(c) While the electron is traveling, it is simultaneously in the gravitational field of the Earth. Explain why it is legitimate to neglect the influence of the gravitational field; include a calculation in your explanation.

5. (10 points)

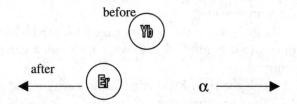

An α-particle is emitted with velocity 1.7×10^7 m/s to the right by an initially motionless Ytterbium nucleus with mass number $A = 173$, as shown above. Ytterbium has atomic number $Z = 70$.

(a) Determine the mass number A and atomic number Z of the resulting Erbium nucleus.

(b) Calculate the magnitude of the momentum of the resulting Erbium nucleus.

Radiation beam incident on Erbium nucleus initially moving to the left

After the decay, a beam of radiation of frequency 450 Hz is incident upon the Erbium nucleus opposite the direction of motion. Each photon in the beam is reflected back by the Erbium. Approximately 10^{25} photons strike the nucleus each second.

(c) Calculate how much time the light would have to be incident on the nucleus to stop its motion.

GO ON TO THE NEXT PAGE

6. (10 points)

An accident occurred in a parking lot. Police lieutenants examined the scene of this accident and found that one of the involved cars, which just about came to a stop before the collision, made skid marks 80 m long. The coefficient of kinetic friction between the rubber tires and the dry pavement was about 0.80; a typical car has a mass of 700 kg.

(a) Draw a free-body diagram representing the forces on the car as it skidded. Assume the car was moving to the right.
(b) Assuming a level road, estimate how fast this car was initially traveling.
(c) The driver was cited for speeding. Explain why.
(d) In court, the driver explains that his car is particularly heavy—it has a mass not of 700 kg, but of 2100 kg. Thus, he suggests that calculations showing he was speeding are invalid. Explain whether the driver's argument makes sense.

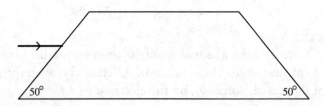

7. (10 points)

A light ray traveling in air is incident upon a glass ($n = 1.5$) trapezoid as shown above.

(a) Draw and label the normal to the surface at the point at which the light enters the glass.
(b) Determine the angle θ_2 at which the light enters the glass; draw the transmitted ray on the above diagram and label θ_2 on your drawing.
(c) Will the light ray leave the glass out of the bottom edge or the right-hand edge?
   — If the light leaves through the bottom edge, at what angle will it enter the air? Show all calculations and draw the transmitted ray.
   — If the light leaves through the right-hand edge, explain why it will not leave through the bottom edge. Show all calculations and draw the path of the ray through the glass.

STOP. End of Physics B—Practice Exam—Free-Response Questions

Physics B—Practice Exam—Multiple-Choice Solutions

1. **B**—Only the vertical motion affects the time in the air. Set up kinematics, with up positive:

    $v_o = 25 \sin 53° \text{ m/s} = 20 \text{ m/s}$.

    $v_f = ?$

    $\Delta x = -25 \text{ m}$.

    $a = -10 \text{ m/s}^2$.

    $t = $ *what we're looking for.*

    Use *** ($v_f^2 = v_o^2 + 2a\Delta x$), then * ($v_f = v_o + at$):

    $v_f^2 = (20 \text{ m/s})^2 + 2(-10 \text{ m/s}^2)(-25 \text{ m})$.

    $v_f^2 = (400) + (500)$.

    $v_f = -30 \text{ m/s}$, negative because ball moves down at the end.

    So * says $-30 \text{ m/s} = 20 \text{ m/s} + (-10)t$; $t = 5.0 \text{ s}$.

2. **A**—Set up a kinematics chart with down positive:

    $v_o = 0$ because dropped.

    $v_f = v$, given.

    $\Delta x = h$.

    $a = g$.

    $t = $ unknown.

    Using *** ($v_f^2 = v_o^2 + 2a\Delta x$) with the variables shown above, $v^2 = 2gh$ so $v = \sqrt{2gh}$. We want to know where the speed is $v/2$, or $\sqrt{2gh}/2$. Reset the kinematics chart and solve:

    $v_o = 0$ because ball dropped.

    $v_f = \sqrt{2gh}/2$.

    $\Delta x = $ *what we're looking for.*

    $a = g$.

    $t = $ unknown.

    Using ***, $2gh/2 = 2g\Delta x$; $\Delta x = h/4$.

3. **D**—The cart must slow down as it goes up the hill. Only A, B, and D start with nonzero speed that approaches zero speed after a long time. The speed reaches zero and then the cart speeds up again *in the same direction.* Only D keeps the motion in the same direction after the pause.

4. **B**—Because north is positive, and the velocity is always in the negative section of the graph, the car must move to the south. Because the speed is getting closer to zero, the car slows down.

5. **D**—Acceleration is the slope of the v–t graph. Here, we only want the slope while the car speeds up: $(30 - 10)\text{m/s} / (30 - 20)\text{s} = 2.0 \text{ m/s}^2$.

6. **D**—The free-body diagram looks like this:

    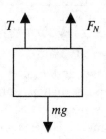

    Because the block is in equilibrium, $F_N + T = mg$. Solving, $F_N = mg - T$.

7. **D**—While in the region between the plates, the positively charged proton is attracted to the negative plate, so bends downward. But after leaving the plates, there is no more force acting on the proton. Thus, the proton continues in motion in a straight line by Newton's first law.

8. **B**—Consider the horizontal and vertical forces separately. The only horizontal forces are the horizontal components of the tensions. Because the block is in equilibrium, these horizontal tensions must be *equal*, meaning only choices B and D can be right. But the ropes can't have equal horizontal AND vertical tensions, otherwise they'd hang at equal angles. So D can't be the correct choice, and B must be.

9. **B**—Tension, friction, and weight are specific forces that can pull or drag an object. But "mass times acceleration" is what we set the sum of all forces equal to—ma is not itself a force.

10. **D**—If the object is rolling straight at constant speed, it is in equilibrium (regardless of its speed), and the net force on it is zero. But zero net force can be obtained either by rightward forces canceling leftward forces, or by no forces acting at all.

11. **C**—Consider the free-body diagram of the person:

F_{scale}

weight

Call up positive, and $F_{scale} - weight = ma$. The scale reading is less than his normal weight, so the acceleration is negative; i.e., *down*. Downward acceleration means either something speeds up while moving downward, or slows down while moving upward.

12. **B**—Work done by gravity depends only on vertical displacement, not on the total distance moved. The displacement of each mass is L. Work is force (here, weight) times this displacement. It takes $3mgL$ of work to raise the bottom mass, but while that's done, the small mass moves down; this means that negative mgL of work was done to lower the small mass. The total work necessary, then, is $3mgL + (-mgL) = 2mgL$.

13. **C**—Speed means the magnitude of the velocity vector, *not* just the vertical velocity; so you can't just use vertical kinematics. You find the final velocity in both directions, and add the vectors. It's easier, though, to use energy. Call point A the top of the table, point B the floor. $PE_A + KE_A = PE_B + KE_B$; $mgh + \frac{1}{2}mv^2 = 0 + \frac{1}{2}mv_B^2$. Cancel the m and solve for v_B, and you get to

$$\sqrt{v^2 + 2gh}$$

14. **D**—Momentum must be conserved in the collision. If block A bounces, it changes its momentum by a larger amount than if it sticks. This means that block B picks up more momentum (and thus more speed) when block A bounces. The mass of the blocks is irrelevant because the comparison here is only between bouncing and not bouncing. So B goes faster in case I regardless of mass.

15. **C**—The period depends only on the mass attached to the spring, not at all on the speed. When block A sticks, more mass is attached, and the period is longer according to

$$T = 2\pi\sqrt{\frac{m}{k}}$$

Which mass is bigger is irrelevant; two blocks will always be heavier than one block.

16. **A**—The relevant equation here is for impulse: $\Delta p = F\Delta t$. The force on the block can be increased either by increasing the momentum change or by decreasing the time of collision. E decreases collision time; B, C, and D all increase the momentum change. But A *increases* the collision time, thereby decreasing the force.

17. **B**—The system has momentum 3 N·s up, and 9 N·s left. The magnitude of the momentum vector is the Pythagorean sum of 3 and 9 N·s. Without calculating, we know the answer must be more than 9 N·s and less than 12 N·s.

18. **D**—Change in momentum is $F\Delta t$. What does that force represent? The *net* force, which here is 2000 N to the left. 2000 N times 20 s is 40,000 N·s, equivalent to 40,000 kg·m/s.

19. **E**—Fields are vectors because they cause forces, and force is a vector. Current is a vector because charge can flow in either direction in a wire. However, electric potential is related (by $PE = qV$) to potential energy, and all forms of energy are scalar quantities.

20. **C**—The velocity of an object in circular motion is NOT constant because the direction of motion changes. The acceleration and net force must be toward the center of the circle. Because the object's velocity vector is tangent to the circle, acceleration and velocity must be perpendicular to each other.

21. **C**—At point A the mass is momentarily at rest, just before it retraces its path back to the bottom. Acceleration is a *change* in velocity. Velocity is zero now, and in a moment will be along the circular path of the pendulum. So acceleration must point in the direction of this velocity change.

22. B—Centripetal acceleration is v^2/r, and is always center *seeking*; here this must be toward the sun. But the problem asks for the net force, which is mass times acceleration.

23. B—The period of a mass on a spring is

$$T = 2\pi\sqrt{\frac{m}{k}}$$

so multiplying the mass by 4 multiplies the period by 2. The total mechanical energy is equal to the potential energy at the maximum displacement. Here, because the amplitude (i.e., the maximum displacement) doesn't change, the total mechanical energy is also unchanged.

24. D—In an orbit, the centripetal force is provided by gravity, $mv^2/r = GMm/r^2$. Solving for the radius of the orbit, $r = GM/v^2$. (Here M is the mass of the moon; the mass of the satellite cancels out.) Doubling the speed, which is squared and in the denominator, divides the radius of the orbit by 4.

25. B—Don't try to calculate the answer by saying $mg = GMm/r^2$! Not only would you have had to memorize the mass of the Earth, but you have no calculator and you only have a minute or so, anyway. So think: the acceleration must be less than 9.8 m/s² because that value is calculated at the surface of the Earth, and the Shuttle is farther from Earth's center than that. But the added height of 300 km is a small fraction (~5%) of the Earth's radius. So the gravitational acceleration will not be THAT much less. The best choice is thus 8.9 m/s². (By the way, acceleration is not zero—if it were, the Shuttle would be moving in a straight line, and not orbiting.)

26. C—Ice is 90% as dense as liquid water. So 90% of its volume is under the water, *displacing* water. Okay, when the ice melts, the new liquid takes up only 90% of the space it originally occupied. This liquid's volume is equal to the amount of volume originally displaced by the ice, so the water level does not rise.

27. C—Use Bernoulli's equation with the flowing fluid. Consider inside and outside the building as the two positions. The y terms are equal, so get rid of them. Call the velocity inside the building zero. We're left with $\frac{1}{2}\rho v_1^2 + P_1 = P_2$, where 2 means inside the building and 1 is outside. So: pressure outside is less than pressure inside. By how much? Well, if you happened to remember that the density of air is about 1 kg/m³, you can make the calculation. But just look at the choices. Ambient pressure doesn't even drop by half when you go to the top of a mountain! So the answer is that pressure drops by a couple of percentage points.

28. B—The free-body diagram of the mass includes the buoyant force AND the normal force up, and mg down. Because the block is in equilibrium, $F_N + F_B = mg$. So the normal force is *less* than mg and certainly not equal to the weight.

29. A—For a fluid of uniform density, the pressure at depth d is $\rho g d$ because of the weight of the water above depth d. But here, because the density of the water gets bigger, the weight of the water is greater. So the pressure should be *greater* than $\rho_o g d$. The pressure is *less* than $\rho g d$, because ρ is the maximum density.

30. D—Don't use the length expansion equation here, because we don't know the expansion coefficient for brass. Rather, just know that the change in an object's length is always a very small fraction of the original length. By

$$T = 2\pi\sqrt{\frac{L}{g}}$$

if the length of the pendulum increases, the period increases as well, eliminating choices (A), (B), and (C). In thermal length expansion, length increases by only a small fraction of the original length . . . there's no possibility of even doubling the pendulum's original length, let alone quadrupling the original length to produce a 4.0 s period. Choice (D) is correct. Try it on your calculator— if the pendulum's length increases to, say, 1.01 m, the period is still 2.0 s.

31. C—The net work done on or by a gas is the area inside the cycle on the *PV* diagram, so the amount of net work is the same in either direction. In *ABCA* more work is done ON the gas than by the gas because there's more area under the curve when the volume compresses than when volume expands. By the same logic, in *ACBA* more work is done BY the gas than on the gas.

32. E—rms speed is given by

$$v_{rms} = \sqrt{\frac{3k_bT}{m}}$$

Use round numbers to make calculations easier. v_{rms} would have to be multiplied by a bit more than 20 to reach escape velocity. So the temperature would have to be multiplied by more than 20^2, or more than 400 times.

33. A—By the ideal gas law, $V = Nk_bT/P$. Because none of the terms on the right change when we discuss oxygen rather than helium, the volume should be the same.

34. C—No time or calculator to do the calculation here, but the answer choices allow for an order of magnitude estimate. A molecule is a few Angstroms, or 10^{-10} m, across. The separation between molecules must be bigger than this. But molecules certainly aren't a millimeter apart from one another! The only choice that makes sense is C.

35. A—B, D, and E are all alternative statements of the second law. C is a consequence of that law—even though energy would be conserved if the putty cooled off and jumped, this action is never observed. A, the correct choice, is a statement of the FIRST law of thermodynamics, which itself is a statement of conservation of energy.

36. B—An electric field exists regardless of the amount of charge placed in it, and regardless of whether any charge at all is placed in it. So both experimenters must measure the same field (though they will measure different forces on their test charges).

37. C—Ah, the Newton's third law question. Whether we're talking about cars or charges, the force on object *A* due to object *B* is always the same as the force on object *B* due to object *A*. Thus the forces described must be equal.

38. A—The charge is in equilibrium, so the horizontal component of the tension must equal the electric force. This horizontal tension is 0.1 N times sin 30° (not cosine because 30° was measured from the *vertical*), or 0.05 N. The electric force is *qE*, where *q* is 0.002 C. So the electric field is 0.050 N/0.002 C . . . reduce the expression by moving the decimal to get 50/2, or 25 N/C.

39. A—In a *metal* sphere, charge is free to move, so the charge ends up on the surface evenly distributed. The electric fields due to each of these charges cancel at the center because for every charge causing a field pointing one way, another charge on the other side of the sphere will cause an equal field pointing the other way. (It's worth knowing that the electric field ANYWHERE inside a conductor is always zero; the justification for this is done in Physics C with Gauss's law.)

40. C—Capacitance of a parallel plate capacitor is $\varepsilon_o A/d$, where *A* is the area of the plates and *d* is the separation between plates. To halve the capacitance, we must halve the area or double the plate separation. The plate separation in the diagram is labeled *c*, so double distance *c*.

41. E—Connected in parallel means that the *voltage* for each capacitor must be the same. The capacitors are identical, so the capacitance of each is the same. Thus, by $Q = CV$, the charge on each must be the same, too.

42. B—Voltage is always the same across parallel resistors. Because the 2.0 Ω resistor is hooked directly to the battery, the voltage across it is 1.0 V. By Ohm's law, $I = V/R = 0.50$ amps.

43. C—The 5 A of current splits into two paths. The path through R_2 takes 2 A, so the path through R_3 takes 3 A. Now use Ohm's law: $V = (3\text{ A})(10\text{ }\Omega) = 30$ V.

44. D—The voltage must stay the same because the battery by definition provides a constant voltage. Closing the switch adds an extra parallel branch to the network of resistors. Adding a parallel resistor REDUCES the total resistance. By Ohm's law, if voltage stays the same and resistance decreases, total current must increase.

45. C—Kirchoff's loop rule says that the sum of voltage changes around a closed circuit loop is zero. When a charge gains or loses voltage, it is also gaining or losing *energy*. The fact that the charge must end up with the same energy it started with is a statement of energy conservation.

46. E—A charge in a uniform magnetic field takes a circular path because the magnetic force changes direction, so it's always perpendicular to the charge's velocity. Here the magnetic force by the right-hand rule is initially down. (Did you think the force on the charge was up? An electron has a negative charge, so the right hand rule gives a direction *opposite* to the direction of the force.)

47. B—Work is force times parallel displacement. Because the magnetic force is always perpendicular to the particle's direction of motion, no work can be done by the magnetic field.

48. B—The magnetic field caused by a straight wire wraps around the wire by the right-hand rule. Consider just little parts of the curvy wire as if the parts were momentarily straight. The part of the wire at the top of the page produces a B field out of the page in the region inside the loop; the parts of the wire at the bottom of the page or on the side of the page also produce B fields out of the page inside the loop. These fields reinforce each other at the center, so the net B field is out of the page.

49. B—The magnetic field produced by I_1 is out of the page at the location of wire 2. The current I_2 is thus moving in a magnetic field; the force on that current-carrying wire is given by ILB and the right-hand rule, which shows a rightward force.

50. D—Lenz's law dictates that the direction of induced current is in the direction of decreasing flux by the right-hand rule. The magnetic field from the bar magnet points out of the north end, which is down. When the magnet gets closer to the loop of wire, the field gets stronger, so flux increases in the down direction; therefore, the direction of decreasing flux is the other way, up; and the right-hand rule gives an initial current counterclockwise. As the magnet falls away from the loop, the field still points down (into the south end), but the flux decreases, so now the current flows clockwise. Only choice D shows counterclockwise current changing to clockwise current.

51. A—You remember the equation for the induced EMF in a moving rectangular loop, $\varepsilon = Blv$. Here l represents the length of the wire that doesn't change within the field; dimension *a* in the diagram. So the answer is either A or C. To find the direction of induced current, use Lenz's law: The field points out of the page, but the flux through the loop is *increasing* as more of the loop enters the field. So, point your right thumb into the page (in the direction of decreasing flux) and curl your fingers; you find the current is clockwise, or left-to-right across the resistor.

52. D—An electric field exerts a force on all charged particles parallel to the field; thus situation (A) would slow the proton down, but not change its direction. Magnetic fields only apply a force if they are perpendicular to a charge's velocity, so (B) and (E) cannot change the proton's direction. A magnetic field perpendicular to motion DOES change a proton's direction, but if a suitable electric field is also included, then the electric force can cancel the magnetic force. So choice D is the only one that must change the direction of the proton.

53. E—The two waves described move through different media (one through the string, one through the air), so must have different speeds. But, because the sound wave was generated by the string wave, the waves have the same frequencies. By $v = \lambda f$, the wavelength must be different for each as well.

54. B—The time between crests is the period of the wave. Period is 1/frequency. So the frequency is 1/(0.2 s), which is 5 Hz.

55. C—Use $v = \lambda f$. The wavelength of the string bass is *twice* the height (because it's a standing wave fixed at both ends), or 4 m. The frequency is $\frac{v}{\lambda} = \frac{200\,\text{m/s}}{4\,\text{m}} = 50\,\text{Hz}$. When the wave changes materials—from the string to the air—the frequency is unchanged.

56. A—Frequency determines color. Think about seeing underwater. You see the same colors in clear water that you do in air. But underwater, the speed of the light waves is slower, and the wavelength is shorter. So if wavelength or wave speed determined color, you would see different colors underwater. (Amplitude determines brightness, and the medium through which the wave travels just determines wave speed.)

57. **C**—Light waves change phase by 180° when they are reflected off a material with higher index of refraction; this only happens in the diagram when light reflects off of the bottom surface.

58. **C**—For simplicity's sake, let's assume that the light is not refracted through huge angles, so we can say that $x = m\lambda L/d$, where x is the distance between spots. d is the distance between lines in the grating. So if d is cut in half, x is doubled; the spots are twice as far apart. The size of the spots themselves should not change.

59. **C**—$E = hf$, so the higher frequency light also has higher energy per photon. But wavelength varies inversely with frequency by $v = \lambda f$, so bigger frequency means smaller wavelength. You should know either that red light has the longest wavelength, or that violet light has the highest frequency.

60. **E**—(A) is wrong—light *slows down* in glass. (C) is wrong—waves do *not* change frequency when changing media. (D) is wrong—light doesn't change direction at all when hitting an interface directly from the normal. Choice (B) is wrong because, while some red light does reflect at each interface, this doesn't help explain anything about seeing light upon exiting the prism. However, blue light does experiences a larger index of refraction than red, so it bends farther from normal when it hits an interface at an angle.

61. **A**—The light from the building that is very far away will be focused at the focal point of the lens. The first part of the problem lets you figure out the focal point using the lensmaker's equation, $1/f = 1/30$ cm $+ 1/20$ cm, so $f = 60/5 = 12$ cm.

62. **D**—All can produce virtual images. However, the converging instruments (the concave mirror and convex lens) can only show virtual images if the object is inside the focal point; these images are larger than the object itself. (Think of a magnifying glass and the inside of a spoon.) The convex mirror and concave lens are diverging instruments, which can ONLY produce smaller, upright images. (Think of a Christmas tree ornament.)

63. **B**—All mirror rays through the center reflect right back through the center.

64. **A**—The frequency of incident photons affects the energy of ejected electrons; the intensity of the beam (number of photons per second) is irrelevant. Thus, the KE vs. intensity graph is simply a horizontal line.

65. **E**—To jump from E_0 to E_1 the electron needs to absorb 4.0 eV of energy. So the necessary wavelength is given by 1240 eV·nm/4.0 eV, or 310 nm. ($E = hc/\lambda$, and $hc = 1240$ eV, as stated on the Constant Sheet.)

66. **D**—The electron can drop from E = 0 to E_0 or to E_1. It can also drop from E_1 to E_0. That makes three different transitions, and three different wavelengths (i.e., colors) of photons.

67. **D**—The stopping voltage is applied to a beam of electrons. Because 1 eV is the energy of 1 electron moved through 1 V of potential difference, these electrons initially had 1 eV of kinetic energy (that was converted into potential energy in order to stop the electrons). These electrons needed 3.5 eV to escape the atom, plus 1 eV for their kinetic energy; this makes a total of 4.5 eV needed from each photon.

68. **B**—In choices A and E, it is light exhibiting wave properties; light is not made up of massive particles. C involves surface waves on the ocean, not the wave behavior of water molecules themselves. Choice D describes Rutherford's experiment, which demonstrated the existence of the atomic nucleus. So choice B is correct: a beam of electrons is a beam of *massive* particles. The experiment described demonstrates both diffraction (bending) and interference, both of which are wave properties.

69. **E**—Only gamma decay keeps atomic mass and number the same, because a gamma particle is just a high energy photon; a photon has neither mass nor charge.

70. **C**—Newton's third law requires the daughter nucleus to experience a force opposite to the force exerted on the alpha particle; thus, the daughter must move backward. Conservation of momentum requires the net momentum to be zero after the decay as well as before; this only happens if the daughter moves backward to cancel the alpha's forward momentum. But energy is a scalar, and has no direction. Conservation of energy can be satisfied no matter which way the particles move.

Physics B—Practice Exam—Free-Response Solutions

Notes on grading your free-response section

For answers that are numerical, or in equation form:

- For each part of the problem, look to see if you got the right answer. If you did, and you showed any reasonable (and correct) work, give yourself full credit for that part. It's okay if you didn't explicitly show EVERY step, as long as some steps are indicated, and you got the right answer. However:

- If you got the WRONG answer, then look to see if you earned partial credit. Give yourself points for each step toward the answer as indicated in the rubrics below. Without the correct answer, you must show each intermediate step explicitly in order to earn the point for that step. (See why it's so important to show your work?)

- If you're off by a decimal place or two, not to worry—you get credit anyway, as long as your approach to the problem was legitimate. This isn't a math test. You're not being evaluated on your rounding and calculator-use skills.

- You do not have to simplify expressions in variables all the way. Square roots in the denominator are fine; fractions in nonsimplified form are fine. As long as you've solved properly for the requested variable, and as long as your answer is algebraically equivalent to the rubric's, you earn credit.

- Wrong, but consistent: Often you need to use the answer to part (a) in order to solve part (b). But you might have the answer to part (a) wrong. If you follow the correct procedure for part (b), plugging in your incorrect answer, then you will usually receive *full credit* for part (b). The major exceptions are when your answer to part (a) is unreasonable (say, a car moving at 10^5 m/s, or a distance between two cars equal to 10^{-100} meters), or when your answer to part (a) makes the rest of the problem trivial or irrelevant.

For answers that require justification:

- Obviously your answer will not match the rubric word-for-word. If the general gist is there, you get credit.

- But the reader is not allowed to interpret for the student. If your response is vague or ambiguous, you will NOT get credit.

- If your response consists of both correct and incorrect parts, you will usually not receive credit. It is not possible to try two answers, hoping that one of them is right. ☺ (See why it's so important to be concise?)

1. (a)

1 pt: It's a frictionless plane, so the only force parallel to the incline is a component of gravity, $mg \sin 30°$.

1 pt: Set this force equal to ma, so $a = g \sin 30 = 5$m/s^2 down the plane.

1 pt: Set up kinematics, with $v_o = 0$, and $\Delta x = 7.0$ m. Solve for time using **, ($\Delta x = v_o t + \frac{1}{2} at^2$).

1 pt: The result is $t = 1.7$ s.

(b)

1 pt: Use the kinematics from part (a), but use * ($v_f = v_o + at$) to get $v_f = 8.5$ m/s.

1 pt: The direction must be down the plane, or 30° below the horizontal.

(c)

1 pt: Only the vertical motion will affect the time to hit the bottom. (This point awarded for any clear attempt to use vertical kinematics.)

1 pt: In the vertical direction, $v_o = (8.5$ m/s)$\sin 30°$; $\Delta x = 4.0$ m and acceleration is $g = 10$ m/s^2.

1 pt: You can use ** ($\Delta x = v_o t + \frac{1}{2} at^2$) and a quadratic to solve for time; it's probably easier to solve for v_f first using *** ($v_f^2 = v_o^2 + 2a \Delta x$). v_f works out to 9.9 m/s. Use * ($v_f = v_o + at$) to get $t = 0.6$ s.

1 pt: But this time is only the time after the package left the incline. To find the total time, the 1.7 s on the plane must be added to get 2.3 s total.

(d)

1 pt: In the horizontal direction, the package maintains constant speed after leaving the plane.

1 pt: So the horizontal distance is the horizontal velocity (equal to 8.5 m/s[cos 30°]) times the 0.6 seconds the package fell.

1 pt: The package lands 4.4 m from the left-hand edge.

(e)

1 pt: The horizontal distance would be increased, but less than doubled.

1 pt: The package speeds up as it falls; therefore, the package doesn't take as much time to fall the additional 4 m as it did to fall the original 4 m. The horizontal velocity doesn't change, so the package can't go as far in this additional time.

2. (a)

1 pt: The total momentum of both carts before collision is $m_A v_A$, because cart B has no speed before collision.

1 pt: The total momentum after collision is $m_A(-v_A') + m_B v_B'$, the negative sign arising because cart A moves backward after collision.

1 pt: So the fraction of momentum conserved is the total momentum after collision divided by the total momentum before collision, $[m_B v_B' - m_A v_A']/[m_A v_A]$.

(b)

1 pt: The total kinetic energy of both carts before collision is $\frac{1}{2}m_A(v_A)^2$.

1 pt: The total kinetic energy after collision is $\frac{1}{2}m_A(v_A')^2 + \frac{1}{2}m_B(v_B')^2$. (The negative sign on v_A' disappears when velocity is squared.)

1 pt: So the fraction of KE conserved is the total KE after collision divided by the total KE before collision, $[\frac{1}{2}m_A(v_A')^2 + \frac{1}{2}m_B(v_B')^2]/[\frac{1}{2}m_A(v_A)^2]$.

(c)

1 pt: Only the expression in part (a) should always be 100%.

1 pt: Momentum is conserved in all collisions, but kinetic energy can be converted into heat and other forms of energy.

(d)

1 pt: The motion detector can be used to measure the speed of cart A before collision.

1 pt: The detector should be placed at the left edge of the track.

1 pt: The speed can be read directly from the detector's output.

1 pt: The motion detector can be used in the same manner to measure the speed of cart A after collision.

1 pt: The stopwatch can be used to measure cart B's speed after collision. Time how long the cart takes to reach the end of the track after the collision.

1 pt: The meter stick can be used to measure the distance cart B traveled in this time.

1 pt: Because the track is frictionless, the cart moves at constant speed equal to the distance traveled over the time elapsed.

3. (a)

1 pt: The buoyant force on this cube is equal to the weight of the water displaced.

1 pt: Half of the block displaces water; half of the block's volume is 0.5 m³.

1 pt: Using the density of water, the mass of 0.5 m³ of water is 500 kg, which has weight 5000 N.

1 pt: Because the cube is in equilibrium, the buoyant force must also equal the weight of the cube. The cube thus has weight 5000 N, and mass 500 kg.

(b)

1 pt: Again, the buoyant force is equal to the weight of the displaced water. This time 0.75 m³ are displaced, making the buoyant force 7500 N.

1 pt: The weight of the mass is 5000 N, so the net force is the up minus down forces, or 2500 N.

(c)

1 pt: When an object is in simple harmonic motion, the net force is equal to kx, where x is the displacement from the equilibrium position.

1 pt: Using the information from part (b), when the displacement from equilibrium is 0.25 m, the net force is 2500 N.

1 pt: So the "spring constant" is 2500 N/0.25 m = 10,000 N/m.

(d)

1 pt: The period of a mass in simple harmonic motion is

$$T = 2\pi\sqrt{\frac{m}{k}}$$

1 pt: Plugging in the mass and spring constant from parts (a) and (c), the period is 1.4 s.

1 pt: Frequency is $1/T$ by definition, so the frequency is 0.71 Hz.

(e)

1 pt: The frequency would be the same.

1 pt: The period of a mass in simple harmonic motion is independent of the amplitude.

1 additional point: For correct units on at least three answers in parts (a), (b), (c), or (d), and no incorrect units.

4. (a)

1 pt: Over a potential difference of 500 V, the electron converts $qV = 8.0 \times 10^{-17}$ J of potential energy into kinetic energy.

1 pt: Set this energy equal to $\frac{1}{2}mv^2$ and solve for v;

1 pt: You get 1.3×10^7 m/s.

(Alternate solution is to approach this part of the problem like part (b); full credit is earned if the solution is correct.)

(b)(i)

1 pt: The electric field due to parallel plates is V/d. This works out to 100 V/0.05 m = 2000 N/C. (d represents the distance between plates.)

1 pt: The force on the electron is $qE = 3.2 \times 10^{-16}$ N.

1 pt: The force is directed upward, toward the positive plate.

(b)(ii)

1 pt: The electron experiences no forces in the horizontal direction, so it maintains a constant horizontal velocity of 1.3×10^7 m/s.

1 pt: Using distance = velocity × time, it takes 3.8×10^{-8} s to travel the 0.5 m between the plates.

(b)(iii)

1 pt: The velocity at point C has both a horizontal and vertical component. The horizontal velocity is still 1.3×10^7 m/s.

1 pt: But in the vertical direction, the electron speeds up from rest. (This point is given for ANY recognition that the vertical direction must be treated separately from the horizontal direction.)

1 pt: Its acceleration is given by the force in part (b)(i) over its mass: $a = 3.2 \times 10^{-16}$ N/9.1 × 10^{-31} kg, = 3.5×10^{14} m/s².

1 pt: Now use a Kinematics Chart to find the vertical speed at point C, with $v_o = 0$ and time given in part (b)(ii); this gives 1.3×10^7 m/s.

1 pt: The speed is the magnitude of the velocity vector. The electron's velocity has components to the right and up; add these using the Pythagorean theorem to get 1.8×10^7 m/s.

(c)

1 pt: The electric force, from part (b)(i), was 10^{-16} N. The gravitational force is $mg = 10^{-30}$ N.

1 pt: So the gravitational force is many orders of magnitude smaller than the electric force, and is, thus, negligible.

5. (a)

1 pt: The alpha particle consists of two protons and two neutrons. So the atomic number Z of the Erbium must be two fewer than that of the Ytterbium, $Z = 68$.

1 pt: The mass number A includes both protons *and* neutrons, so must be four fewer in the Erbium than in the Ytterbium, $A = 169$.

(b)

1 pt: Initial momentum of the whole system was zero; so the alpha and the Erbium must have equal and opposite momenta after decay.

1 pt: The alpha's momentum is (4 amu)(1.7 × 10^7 m/s) = 1.4×10^8 amu·m/s (or 1.3×10^{-19} kg·m/s); the Erbium has this same amount of momentum.

(c)

1 pt: The momentum p of each incident photon is given by $E = pc$, where E is the energy of the photon, equal to hc/λ. Set the two expressions for energy equal to each other.

1 pt: It is found that the momentum of each photon is $p = h/\lambda = (6.6 \times 10^{-34}$ J·s)/ $(450 \times 10^{-9}$ m) = 1.5×10^{-27} N·s. (Note that this answer is in standard units because we plugged in h and λ in standard units.)

1 pt: Each photon is reflected backward, so each photon *changes* momentum by 1.5×10^{-27} N·s − (−1.5 × 10^{-27} N·s) = 3.0×10^{-27} N·s.

1 pt: The nucleus has 1.3×10^{-19} N·s of momentum to start with, as found in (b); this point is awarded for both recognizing to use the momentum from part (b) *and* for putting that momentum in standard units.

1 pt: The nucleus loses 3.0×10^{-27} N·s with each photon; so divide to find that 4.3×10^7 photons will bring the nucleus to rest.

1 pt: Dividing this by the 10^{25} photons each second, only 4×10^{-18} seconds are required to stop the nucleus.

6. (a)

1 pt: The normal force acts up and the weight of the car acts down; friction acts to the left, opposite velocity.

1 pt: There are no forces acting to the right.

(b)

1 pt: Here, the normal force is equal to the car's weight, *mg*.

1 pt: The force of friction is μF_N, which is μmg.

1 pt: The net work is done by friction and is equal to the force of friction times the 80 m displacement. Net work done on the car is thus μmg(80 m).

1 pt: Net work is equal to change in kinetic energy . . . because the car comes to rest, the change in KE is equal to the KE when the car was initially moving.

1 pt: So μmg(80 m) $= \frac{1}{2}mv^2$, and $v = 36$ m/s. (Alternate solution: Use $F_{net} = ma$, where F_{net} is the force of friction $= \mu mg$. Solve for acceleration, then use kinematics with final velocity = zero and displacement = 80 m. The answer should still work out to 36 m/s.)

(c)

1 pt: 36 m/s is something like 80 miles per hour . . . in a parking lot?!?!

(d)

1 pt: No, the driver's argument does *not* make sense.

1 pt: In the equation to determine the initial speed of the car, as shown in part (b), the mass cancels out (because it shows up both in the friction force and in the kinetic energy term). So this calculation is valid for any mass car.

7.

(a)

1 pt: The normal to the surface is, by definition, *perpendicular* to the surface, as shown in the diagram above.

(b)

1 pt: Snell's law is used to determine the angle of transmission: $n_1 \sin \theta_1 = n_2 \sin \theta_2$.

1 pt: Here $n_1 = 1.0$ (for air) and $n_2 = 1.5$.

1 pt: The angle of incidence θ_1 must be measured to the normal.

1 pt: From geometry θ_1 is found to be 40°. So the transmission angle is 25°.

1 pt: This angle must be labeled with respect to the normal, as shown in the diagram above.

(c)

1 pt: The light will leave through the right-hand edge.

1 pt: Using geometry, the angle of incidence of the light on the bottom surface can be found. In the above diagram, the angle labeled *A* is 115° because it is the 90° angle to the normal plus the 25° angle of transmission. So angle *B* must be 15° because there are 180° total in a triangle. Therefore, the angle of incidence of the light ray to the bottom surface is 75°.

1 pt: Using Snell's law, the angle of transmission θ_2 is undefined, so total internal reflection must occur. (Or, the critical angle for total internal reflection between glass and air is $\sin \theta_c = 1/1.5$, so $\theta_c = 42°$; 75° is thus bigger than the critical angle, so total internal reflection occurs.)

1 pt: For a diagram showing reflection (with an angle of reflection approximately equal to the angle of incidence) at the bottom surface.

AP Physics B
Full Exam Scoring

Multiple Choice: Number Correct_____(70 max)

Free Response: Question 1_____(15 max)
                Question 2_____(15 max)
                Question 3_____(15 max)
                Question 4_____(15 max)
                Question 5_____(10 max)
                Question 6_____(10 max)
                Question 7_____(10 max)

Total Free Response_____(90 max)

$1.286 \times$ Multiple Choice + Free Response = Raw Score_____(180 max)

119–180	5
94–118	4
65–93	3
52–64	2
0–51	1

Appendices

CONSTANTS

1 amu	u	1.7×10^{-27} kg
mass of proton	m_p	1.7×10^{-27} kg
mass of neutron	m_n	1.7×10^{-27} kg
mass of electron	m_e	9.1×10^{-31} kg
charge of proton	e	1.6×10^{-19} C
Avogadro's number	N	6.0×10^{-23} mol^{-1}
Universal gas constant	R	8.3 J/(mol·K)
Boltzmann's constant	k_B	1.4×10^{-23} J/K
Speed of light	c	3.0×10^{8} m/s
Planck's constant	h	6.6×10^{-34} J·s
	h	4.1×10^{-15} eV·s
Planck's constant · speed of light	hc	1.99×10^{-25} J·m
	hc	1.24×10^{3} eV·nm
Permittivity of free space	ε_o	8.9×10^{-12} C^2/N·m^2
Coulomb's law constant	k	9.0×10^{9} N·m^2/C^2
Permeability of free space	μ_o	$4\pi \times 10^{-7}$ T·m/A
Universal gravitation constant	G	6.7×10^{-11} N·m^2/kg^2
Earth's free fall acceleration	g	9.8 m/s^2
1 atmosphere of pressure	atm	1.0×10^{5} Pa
1 electron-volt	eV	1.6×10^{-19} J

PHYSICS B EQUATIONS

Read Chapter 6 about memorizing equations for more help with learning not only what the equations say but also what they mean.

Remember, your textbook might use slightly different symbols.

NEWTONIAN MECHANICS

$$v_f = v_o + at$$

$$x - x_0 = v_0 t + \tfrac{1}{2} at^2$$

$$v_f^2 = v_0^2 + 2a(x - x_0)$$

$$F_{net} = ma$$

$$F_f = \mu F_N$$

$$a_c = \frac{v^2}{r}$$

$$\tau = F \cdot d$$

$$p = mv$$

$$I = \Delta p = F \cdot \Delta t$$

$$KE = \tfrac{1}{2} mv^2$$

$$PE_g = mgh$$

$$W = F \cdot \Delta x$$

$$P = \frac{W}{\Delta t}$$

$$P = F \cdot v$$

$$F = -kx$$

$$PE_s = \tfrac{1}{2} kx^2$$

$$T = 2\pi \sqrt{\frac{m}{k}}$$

$$T = 2\pi \sqrt{\frac{L}{g}}$$

$$T = \frac{1}{f}$$

$$F_G = G \frac{m_1 m_2}{r^2}$$

$$PE_G = G \frac{m_1 m_2}{r}$$

ELECTRICITY AND MAGNETISM

$$F = \frac{1}{4\pi\varepsilon_0}\frac{q_1 q_2}{r^2}$$

$$F = qE$$

$$PE_E = qV = \frac{1}{4\pi\varepsilon_0}\frac{q_1 q_2}{r}$$

$$E = \frac{V}{d}$$

$$V = \frac{1}{4\pi\varepsilon_0}\sum\frac{q_i}{r_i}$$

$$Q = CV$$

$$C = \varepsilon_0 \frac{A}{d}$$

$$PE_C = \frac{1}{2}CV^2$$

$$I = \frac{\Delta Q}{\Delta t}$$

$$R = \rho\frac{L}{A}$$

$$V = IR$$

$$P = IV$$

$$C_p = \sum C_i$$

$$\frac{1}{C_s} = \sum\frac{1}{C_i}$$

$$R_s = \sum R_i$$

$$\frac{1}{R_p} = \sum\frac{1}{R_i}$$

$$F = qvB\sin\theta$$

$$F = ILB\sin\theta$$

$$B = \frac{\mu_0 I}{2\pi r}$$

$$\phi_m = BA\cos\theta$$

$$\varepsilon = -N\frac{\Delta\phi}{\Delta t}$$

$$\varepsilon = Blv$$

FLUID MECHANICS AND THERMAL PHYSICS

$$P = P_0 + \rho g h$$

$$F_b = \rho V g$$

$$A_1 v_1 = A_2 v_2$$

$$P_1 + \rho g y_1 + \tfrac{1}{2}\rho v_1^2 = P_2 + \rho g y_2 + \tfrac{1}{2}\rho v_2^2$$

$$\Delta L = \alpha L_0 \Delta T$$

$$H = \frac{kA\Delta T}{L}$$

$$P = \frac{F}{A}$$

$$PV = nRT$$

$$KE_{avg} = \frac{3}{2}k_B T$$

$$v_{rms} = \sqrt{\frac{3k_B T}{m}}$$

$$W = -p\Delta V$$

$$\Delta U = Q + W$$

$$e = \frac{W}{Q_H}$$

$$e_{ideal} = \frac{T_H - T_C}{T_H}$$

WAVES AND OPTICS

$$v = \lambda f$$

$$\frac{1}{f} = \frac{1}{d_o} + \frac{1}{d_i}$$

$$d \sin \theta = m\lambda$$

$$n = \frac{c}{v}$$

$$n_1 \cdot \sin \theta_1 = n_2 \cdot \sin \theta_2$$

$$m = \frac{h_i}{h_o} = -\frac{d_i}{d_0}$$

$$x = \frac{m\lambda L}{d}$$

$$\sin \theta_c = \frac{n_2}{n_1}$$

$$f = \frac{r}{2}$$

ATOMIC AND NUCLEAR PHYSICS

$$E = hf = pc$$

$$\lambda = \frac{h}{p}$$

$$\mathrm{KE} = hf - W$$

$$E = (\Delta m)c^2$$

FOUR-MINUTE DRILL PROMPTS

The lists that follow are designed to help you study equations. Each prompt refers to a specific equation on the AP Equations Sheet (we've listed the prompts in the same order in which the equations appear on the Equations Sheet). So, for example, the prompt "Net force" refers to the equation, "$F_{net} = ma$."

There are several ways to use these prompts. First, you can use them as a self-test: for each prompt, write down the corresponding equation on a separate sheet of paper. Then check the equations you wrote down against the AP Equations Sheet to see if you got any wrong. You can also use these prompts when you study with a friend: have your friend read the prompts to you, and you respond by reciting the appropriate equation. Try to go through the list as fast as possible without making a mistake. Last, your physics teacher can use these prompts to lead your class through a Four-Minute Drill, which is an activity we describe in Chapter 6.

Newtonian Mechanics

1st kinematics equation
2nd kinematics equation
3rd kinematics equation
Net force
Force of friction
Centripetal acceleration
Torque
Momentum
Impulse
Kinetic energy
Potential energy due to gravity (near Earth)
Work
Power
Power—alternate expression

Force of a spring (The negative sign reminds you that the spring force is a restoring force, always acting toward the equilibrium point.)
Potential energy of a spring
Period of a mass on a spring
Period of a pendulum
Period in terms of frequency
Gravitational force between two massive objects
Gravitational potential energy between two massive objects. (Don't use unless an object is far from a planet's surface.)

Electricity and Magnetism

Electric force between two point charges—remember, $\dfrac{1}{4\pi\varepsilon_o}$ is just the Coulomb's law constant, $k = 9.0 \times 10^9$ N·m²/C²!!!
Electric force in terms of electric field
Potential energy in terms of potential, and then potential energy between two point charges (this line on the equation sheet really has two different equations. PE = qV is always valid, but PE = kqq/r is only valid between two point charges.)
Uniform electric field between two parallel plates
The potential at some point due to surrounding point charges
Definition of capacitance
Capacitance of a parallel plate capacitor
Energy stored on a capacitor
Definition of current
Resistance of a wire
Ohm's law
Power in an electrical circuit
How to add parallel capacitors
How to add series capacitors
How to add series resistors
How to add parallel resistors
Magnetic force on a charge
Magnetic force on a wire
Magnetic *field* (*not force*) due to a long, straight wire

Magnetic flux

Induced EMF

Induced EMF in the special case of a rectangular wire with constant speed

Fluid Mechanics and Thermal Physics

Pressure in a static fluid column

Buoyant force—Archimedes' principle

Continuity equation

Bernoulli's equation

Thermal expansion

Rate of heat transfer

Definition of pressure

Ideal gas law

Average kinetic energy per molecule in an ideal gas

rms speed of the molecules of an ideal gas

Work done on a gas in an isobaric process (This reminds you that work done on a gas in *any* process can be found by the area under the *PV* graph.)

First law of thermodynamics

Definition of efficiency

Efficiency of an ideal heat engine

Waves and Optics

Velocity of a wave

Definition of the index of refraction

Snell's law

How to find the critical angle for total internal reflection

Lensmaker's equation / Mirror equation (They're the same thing.)

Magnification

Focal length of a spherical mirror

Position of constructive interference points for light passing through slits

Position of constructive interference points for light passing through slits if the angle to the screen is small

Atomic and Nuclear Physics

Energy of a photon—the second expression gives energy in terms of a photon's momentum

Kinetic energy of an electron ejected from a metal surface, where W is the work function

De Broglie wavelength

Conversion between mass and energy

WEB SITES

The Internet offers some great resources for preparing for the AP Physics exam.

- Your textbook may have an associated Web site . . . if so, check it out!
- All *kinds* of stuff associated with the AP Physics B program is contained in Dolores Gende's Web site, http://apphysicsb.homestead.com/. This is the best of the unofficial AP physics sites out there.
- Of course, the official site of the College Board, www.collegeboard.org, has administrative information and test-taking hints, as well as contact information for the organization that actually is in charge of the exam.
- Did you enjoy your first taste of physics? If so, you can try your hand at physics debating. The United States Association for Young Physicists Tournaments hosts a national tournament that consists of "physics fights," or debates, over experimental research projects. Check out www.usaypt.org for details.
- The author writes the country's leading physics teaching blog, available at jacobsphysics.blogspot.com. Students and teachers can obtain and share ideas at this site.

absolute pressure—the total pressure of a fluid at the bottom of a column; equal to the pressure acting on the top of the column plus the pressure caused by the fluid in the column

acceleration—the change in an object's velocity divided by the time it took to make that change; equal to the derivative (slope) of an object's velocity–time function

adiabatic process—a process during which no heat flows into or out of the system

alpha particle—two protons and two neutrons stuck together

amplitude—the maximum displacement from the equilibrium position during a cycle of periodic motion; also, the height of a wave

angular momentum—the amount of effort it would take to make a rotating object stop spinning

antineutrino—a subatomic particle; see "neutrino"

antinode—point on a standing wave where the wave oscillates with maximum amplitude

Archimedes' principle—the buoyant force on an object is equal to the weight of the fluid displaced by that object

atom—the fundamental unit of matter; includes protons and neutrons in a small nucleus, surrounded by electrons

atomic mass unit (amu)—the mass of a proton; also the mass of a neutron

atomic number—the number of protons in an atom's nucleus

average speed—the distance an object travels divided by the time it took to travel that distance (see "speed")

beats—rhythmic interference that occurs when two notes of unequal but close frequencies are played

beta particle—an electron or a positron

buoyant force—the upward force exerted by a fluid on an object that is wholly or partially submerged in that fluid

capacitor—a charge-storage device, often used in circuits

centrifugal force—a made-up force; when discussing circular motion, only talk about "centripetal" forces

centripetal force—the force keeping an object in uniform circular motion

coefficient of friction—the ratio of the friction force to the normal force. The coefficient of static friction is used when an object has no velocity relative to the surface it is in contact with; the coefficient of kinetic friction is used for a moving object.

concave lens—a translucent object that makes the light rays passing through it diverge

conservative force—a force that acts on an object without causing the dissipation of that object's energy in the form of heat

constructive interference—the overlap of two waves such that their peaks line up

convex lens—a translucent object that makes the light rays passing through it converge

critical angle—the angle past which rays cannot be transmitted from one material to another

current—the flow of positive charge in a circuit; the amount of charge passing a given point per unit time

cutoff frequency—the minimum frequency of light that, when absorbed, is capable of making an atom eject an electron

cycle—one repetition of periodic motion

daughter nucleus—the nucleus left over after an atom undergoes nuclear decay

de Broglie wavelength—the wavelength of a moving massive particle

density—the mass of an object divided by that object's volume

destructive interference—the overlap of two waves so that the peaks of one wave line up with the troughs of the other

dipole—something, usually a set of charges, with two nonidentical ends

direction—the orientation of a vector

displacement—a vector quantity describing how far an object moved

Doppler effect—the apparent change in a wave's frequency that you observe whenever the source of the wave is moving toward or away from you

elastic collision—a collision in which kinetic energy is conserved

electric field—a property of a region of space that affects charged objects in that particular region

electric flux—the amount of electric field that penetrates a certain area

electric potential—potential energy provided by an electric field per unit charge

electromagnetic induction—the production of a current by a changing magnetic field

electron—a subatomic particle that carries a negative charge

electron-volt—a unit of energy equal to the amount of energy needed to change the potential of an electron by one volt

energy—the ability to do work

entropy—a measure of disorder

equilibrium—when the net force and net torque on an object equal zero

equipotential lines—lines that illustrate every point at which a charged particle would experience a given potential

field—a property of a region of space that can affect objects found in that particular region

first law of thermodynamics—the change in the internal energy of a system equals the heat added to the system plus the work done on the system

flowing fluid—a fluid that's moving

free-body diagram—a picture that represents one or more objects, along with the forces acting on those objects

frequency—the number of cycles per second of periodic motion; also, the number of wavelengths of a wave passing a certain point per second

friction—a force acting parallel to two surfaces in contact; if an object moves, the friction force always acts opposite the direction of motion

fulcrum—the point about which an object rotates

fundamental frequency—the frequency of the simplest standing wave

g—the acceleration due to gravity near the Earth's surface, about 10 m/s^2

gamma particle—a photon

gauge pressure—the pressure of a fluid at the bottom of a column due only to the fluid in the column

ground state energy—the lowest energy level of an atom

heat—a type of energy (related to molecular vibrations) that can be transferred from one object to another

heat engine—a system in which heat is added to a gas contained in a cylinder with a moveable piston; when the gas heats up, it expands, doing work by moving the piston up

impulse—the change in an object's momentum

index of refraction—a number that describes by how much light slows down when it passes through a certain material

induced EMF—the potential difference created by a changing magnetic flux that causes a current to flow in a wire; EMF stands for "electro-motive force," but the units of EMF are *volts*.

inductance—the property of an inductor that describes how good it is at resisting changes in current in a circuit

inductor—a coil in a circuit that makes use of induced EMF to resist changes in current in the circuit

inelastic collision—a collision in which kinetic energy is not conserved, as opposed to an elastic collision, in which the total kinetic energy of all objects is the same before and after the collision

inertia—the tendency for a massive object to resist a change in its velocity

internal energy—the sum of the kinetic energies of each molecule of a substance

ion—an electrically charged atom or molecule

ionization energy—the minimum amount of energy needed for an electron to escape an atom

isobaric process—a process during which the pressure of the system remains the same

isochoric process—a process during which the volume of the system remains the same

isotherm—a curve on a *PV* diagram for which the temperature is constant

isothermal process—a process during which the temperature of the system remains the same

isotope—an atom with the same atomic number as another atom but a different atomic mass

kinetic energy—energy of motion

Kirchoff's laws—in a circuit, (1) at any junction, the current entering equals the current leaving; (2) the sum of the voltages around a closed loop is zero

Lenz's law—the direction of the current induced by a changing magnetic flux creates a magnetic field that opposes the change in flux

longitudinal wave—when particles move parallel to the direction of a wave's motion

magnetic field—a property of a region of space that causes magnets and moving charges to experience a force

magnetic flux—the amount of magnetic field that penetrates an area

magnitude—how much of a quantity is present; see "scalar" and "vector"

mass defect—the amount of mass that is destroyed and converted into kinetic energy in a nuclear decay process

mass number—the number of protons plus neutrons in an atom's nucleus

mass spectrometer—a device used to determine the mass of a particle

moment of inertia—the rotational equivalent of mass

momentum—the amount of "oomph" an object has in a collision, equal to an object's mass multiplied by that object's velocity

net force—the vector sum of all the forces acting on an object

neutrino—A subatomic particle emitted during beta decay that affects only the kinetic energy of the products of the decay process

neutron—a subatomic particle found in an atom's nucleus that has no electric charge

node—point on a standing wave where the medium through which the wave is propagating does not move

normal force—a force that acts perpendicular to the surface on which an object rests

nucleus—the small, dense core of an atom, made of protons and neutrons

oscillation—motion of an object that regularly repeats itself over the same path

parallel—the arrangement of elements in a circuit so that the charge that flows through one element does not flow through the others

Pascal's principle—if a force is applied somewhere on a container holding a fluid, the pressure increases everywhere in the fluid, not just where the force is applied

peak—a high point on a wave

perfectly inelastic collision—a collision in which the colliding objects stick together after impact

period—the time it takes for an object to pass through one cycle of periodic motion; also, the time for a wave to propagate by a distance of one wavelength

photoelectric effect—energy in the form of light can cause an atom to eject one of its electrons, but only if the frequency of the light is above a certain value

photon—a particle of light; a photon has no mass

plane mirror—a flat, reflective surface

positron—like an electron, but with a positive charge

potential energy—energy of position

power—the amount of work done divided by the time it took to do that work; also, in a circuit, equal to the product of the current flowing through a resistor and the voltage drop across that resistor

pressure—the amount of force applied to a surface divided by the area of that surface

principal axis—the imaginary line running through the middle of a spherical mirror or a lens

principle of continuity—the volume flow rate is equal at all points within an isolated stream of fluid

proton—a subatomic particle found in an atom's nucleus that carries a positive charge

PV **diagram**—a graph of a gas's pressure versus its volume

real image—an image created by a mirror or a lens that is upside-down and can be projected onto a screen

refrigerator—like a heat engine, only work is done to remove heat

resistance—a property of a circuit that resists the flow of current

resistor—something put in a circuit to increase its resistance

restoring force—a force that restores an oscillating object to its equilibrium position

scalar—a quantity that has a magnitude but no direction

second law of thermodynamics—heat flows naturally from a hot object to a cold object but not from cold to hot; equivalently, the entropy of a system cannot decrease unless work is done on that system

series—the arrangement of elements in a circuit so that they are connected in a line, one after the other

specific gravity—the ratio of a substance's density to the density of water

speed—the magnitude of an object's velocity

spherical mirror—a curved, reflective surface

standing wave—a wave that, when observed, appears to have peaks and troughs that don't move

static fluid—a fluid that isn't flowing

temperature—a quantity related to the average kinetic energy per molecule of a substance

tension—a force applied by a rope or string

thermal expansion—enlargement of an object that is heated

time constant—a value related to how long it takes to charge or discharge a capacitor

torque—the application of a force at some distance from a fulcrum; if the net torque on an object isn't zero, the object's rotational velocity will change

total internal reflection—the reflection of light off a surface that occurs when the light cannot pass from a medium with a high index of refraction to one with a low index of refraction

transverse wave—occurs when the particles in a wave move perpendicular to the direction of the wave's motion

trough—a low point on a wave

vector—a quantity that has both magnitude and direction

velocity—how fast an object's displacement changes; equal to the derivative (slope) of an object's position–time function

virtual image—an image created by a mirror or lens that is upright and cannot be projected onto a screen

volume flow rate—the volume of fluid flowing past a point per second

wave—a rhythmic up-and-down or side-to-side motion that moves through a material at constant speed

wavelength—a wave's peak-to-peak or trough-to-trough distance

wave pulse—when a single peak travels through a medium

weight—the force due to gravity; equal to the mass of an object times g, the gravitational field

work—the product of the distance an object travels and the components of the force acting on that object directed parallel to the object's direction of motion

work-energy theorem—the net work done on an object equals that object's change in kinetic energy

work function—the minimum amount of energy needed for an electron to be ejected from the surface of a metal

BIBLIOGRAPHY

Your AP physics textbook may have seemed difficult to read early in the year. But now that you have heard lectures, solved problems, and read our guide, try reading your text again—you'll be amazed at how much more clear the text has become.

If you'd like to look at another textbook, these are a few that we recommend:

- Giancoli, Douglas C. (2004). *Physics* (6th ed.). New York: Prentice Hall.
- Cutnell, J. D., Johnson, K. W. (2009). *Physics* (8th ed.). New York: Wiley.
- Jones, Edwin R., Childers, R. L. (2000). *Contemporary College Physics* (3rd ed.). New York: McGraw-Hill.

You might also find this book helpful:

- Hewitt, P. G. (2009). *Conceptual Physics* (11th ed.). San Francisco: Addison Wesley. (Hewitt's is the classic text for readable, nonmathematical expositions of physics principles. If you are having trouble seeing the meaning behind the mathematics, check out this book.)

Just for fun, we also recommend these books . . . they might not help you too much for the AP exam, but they're great reads.

- Feynman, R. (1997). *Surely You're Joking, Mr. Feynman!* New York: W. W. Norton. (Collected stories of the 20th century's most charismatic physicist. If you ever thought that physicists were a bunch of stuffy nerds without personality, you should definitely read this book. One of our all-time favorites!)
- Hawking, S. (1998). *A Brief History of Time.* New York: Bantam. (The canonical introduction to cosmology at a layperson's level.)
- Lederman, L. (1993). *The God Particle.* New York: Dell. (Written by a Nobel Prize–winning experimental physicist, this book not only discusses what kinds of strange subatomic particles exist, but goes through the amazing and interesting details of how these particles are discovered.)
- Walker, J. (2007). *The Flying Circus of Physics* (2nd ed.). New Jersey: Wiley. (This book provides numerous conceptual explanations of physics phenomena that you have observed. The classic "Physics of the world around you" book.)

NOTES